Greece—a Jewish History

Greece

A JEWISH HISTORY

K. E. Fleming

PRINCETON UNIVERSITY PRESS

PRINCETON AND OXFORD

Copyright © 2008 by Princeton University Press
Published by Princeton University Press, 41 William Street,
Princeton, New Jersey 08540
In the United Kingdom: Princeton University Press, 3 Market Place,
Woodstock, Oxfordshire OX20 1SY

Library of Congress Cataloging-in-Publication Data

Fleming, K. E. (Katherine Elizabeth), 1965–
Greece—a Jewish history / K. E. Fleming.
p.   cm.
Includes bibliographical references and index.
ISBN-13: 978-0-691-10272-6 (hardcover : alk. paper)
1. Jews—Greece—History—20th century.
2. Jews—Greece—History—21st century.
3. Greece—Ethnic relations. I. Title.
DS135.G7F54   2008
949.5′004924—dc22
ISBN 978-0-691-10272-6

British Library Cataloging-in-Publication Data is available

This book has been composed in Sabon

Printed on acid-free paper. ∞

press.princeton.edu

Printed in the United States of America

1   3   5   7   9   10   8   6   4   2

In memory of

Cyril S. Arvanitis
(1926–2004)

and

Hyman Genee
(1922–2006)

οὐκ ἔνι Ἰουδαῖος οὐδὲ Ἕλλην

CONTENTS

ILLUSTRATIONS

ACKNOWLEDGMENTS

The members of the congregation of the Kehila Kedosha Janina, Broome Street, New York, were indispensable to this book and, indeed, the daily life of my family over the course of many years. In particular, Marcia Ikonomopoulos, director of the synagogue's museum, was extraordinarily generous with her time, knowledge, contacts, and resources. Many of the photographs that appear in this book, and the source material for its introductory chapter, were given to me by her. The Jewish Museum of Greece and its director, Zanet Battinou, are also due deep thanks. Several of the photographs that appear in this book are courtesy of the museum's photographic archive; I am grateful for permission to use them, and also for the museum's gracious assistance with producing many of them in record time. I thank Yad Vashem's film and photo archive for permission to reproduce materials from its collection; I am particularly grateful to archivist Naama Shmayovitz, who was a paragon of efficiency. These collections were vital not only for the photographs that appear in this book but even more for the many, many others that do not, yet that gave me a vivid visual image of the history I sought to describe in its core chapters.

I am grateful to Azulai Brothers Productions for permission to reproduce the album cover of Levitros's *Greatest Hits, Volume II*, which appears in chapter 10; I am indebted to Levitros (Levi Mu'alem) himself for speaking to me about his background and interest in Greek music. Warm thanks to Yigal Nizri for tracking down the Azulai Brothers and Levitros, and for doing many other things as well. Thanks also to Cambridge University Press for permission to use the map of the expansion of the Greek state that first appeared in Richard Clogg's *Concise History of Greece* (Cambridge, 1992). Mark Mazower shared documents with me and was generous in the time he spent discussing this project. Niki Kekos graciously did a final read through of my Greek. New York University provided support in the form of a Research Challenge Grant, which underwrote many of the interviews I undertook as part of researching the project; the Robert F. Wagner Graduate School of Public Service provided a semester of leave and significant research support in the form of a Stephen Charney Vladeck Junior Faculty Fellowship during the academic year 2002–3.

Two individuals in particular were central to this project throughout. Hyman Genee, the beloved president of the Kehila Kedosha Janina who died in February 2006, was a living repository of Greek Jewish American history. A superbly gracious, witty, and intelligent personage, he was a constant reminder that I was writing not a past history of a vanished people but the account of an ongoing, if diverse and changing, worldwide community.

Cyril Arvanitis, my self-appointed research assistant, bibliographer, critic, and editor, was actively involved in the nuts and bolts of this project from its outset. His linguistic facility, astonishing intellectual breadth, and patience for detail combined with his constant prodding and encouragement to make the completion of this book come about sooner than it otherwise would have. Only with his death, in December 2004, did I fully realize that he had become much more to me than the modest role in which he'd cast himself suggested.

I miss them both deeply and dedicate this book to their memory.

Greece—a Jewish History

Chapter 1

INTRODUCTION

On October 26, 1914, Thomas Donnelly, justice of the Supreme Court of the State of New York, signed off on docket #4979–1914 C, legally certifying the incorporation of the "Jewish Community of Janina, Inc." A parenthetical note explained the name to the court: "(Janina [is] the name of a town in Greece)." What was being incorporated was a "Greek Jewish group." These notes were deceptively simple. For Janina (Jannina) had only been part of Greece for one year, and the idea of a "Greek Jew" was all but nonexistent—at least in Greece.

In 1914, Jannina was home to one of the oldest Jewish communities in Europe: the so-called Romaniotes, the indigenous, Greek-speaking Jews who had first settled in the south Balkans in the first centuries C.E. At the same time, it was one of Greece's newest towns: just one year before, in 1913, the town, along with most of what today is northern Greece, had been taken from the Ottoman Empire by Greek forces in the second Balkan War. This "town in Greece," like many others, had spent the previous five hundred years as part of the Ottoman Empire.

Hundreds of thousands of the region's inhabitants, Jewish, Christian, and Muslim alike, were profoundly affected by the transition. The late nineteenth and early twentieth centuries saw massive migration from the Ottoman borderlands to the United States. Between 1873 and 1924, almost 550,000 Greeks from the Ottoman Empire and Greece arrived in America.[1] With them were many Jews, both Greek-speaking Romaniotes from around the south Balkan mainland and Sephardim from the city of Salonika—the descendants of the Jews expelled from the Iberian Peninsula starting in the late fifteenth century. Between 1895 and 1906, up to 30 percent of the total population of the northwestern Greek district of Epirus, of which Jannina was the seat, emigrated.[2] The Jewish Community of Janina, Inc. was made up of recent émigrés who were part of this mass movement.

The corporation's nine original directors had come together to create a social and religious home for the burgeoning new Romaniote community of New York's Lower East Side.[3] All recent immigrants, six members of the board were naturalized U.S. citizens while the remaining third were Greek nationals. The new corporation had as its goal "to unite the Jews of Janina for the promotion of their welfare, physically, morally, and intellectually," and to "aid and assist them financially and morally; to cultivate and foster social intercourse among the members and in general to ameliorate their condition."[4]

The founding of the Jewish Community of Janina, Inc. signaled a significant Greek Jewish presence on lower Manhattan—one that grew in the wake of the Balkan Wars (1912–13) and during the interwar period, when there was another, smaller wave of emigration. Of the roughly six thousand Jews living in Jannina at the turn of the century, more than half had departed by 1930, most to the United States.[5]

By the mid-1920s, when America's new "closed door" immigration policy took effect, slowing the growth of the New York Jannina community, affiliate organizations—a burial society and a benevolent foundation—had been established; money was being raised to build a permanent home for the Janniote congregation. By 1926, the burial society (the United Brotherhood of Janina, which resulted that year from the merger of various other organizations) had 250 members, and total holdings valued at just over $12,000. Annual dues averaged $6.60.[6] A constitution was drawn up, officers elected, and fees set according to age. The next year a permanent synagogue, the Kehila Kedosha Janina (holy community of Jannina), was founded. Services were packed. The New York community was so large that the Janina Brotherhood had a Harlem Division and a Downtown Division, each presided over by its own rabbi.[7]

The Kehila Kedosha Janina, at 280 Broome Street on lower Manhattan, was the social and spiritual hub for the thriving group of new Greek Jewish immigrants. While Janniotes made up the congregation's core, the synagogue drew congregants from all over Greece. Over time, in a nice ironic reversal of the conditions that had prevailed for centuries in the south Balkans, Ladino-speaking Sephardim—who in Greece had dominated Romaniotes in number—were incorporated into the group, which to this day follows the Greek/Judeo *minhag* (customary practice). As in Greece, in New York the Romaniote rabbinic establishment interacted with various Sephardic congregations. Close ties were maintained with the Spanish-Portuguese Synagogue at 133 Eldridge Street, for instance; while the Eldridge Street synagogue had a Hebrew school, the Kehila Kedosha Janina did not, so young New York Romaniotes training for Bar Mitzvah would study with the Sephardim.

In early June 1935, a young man from the Jannina community was Bar Mitzvahed at the Spanish-Portuguese Synagogue. "It was a very memorable day for me, my family and friends." The celebration lasted for two days; the following weekend when the young man was called up to the Torah back at the Jannina synagogue another party followed.[8] The young man was Hyman Genee, who until his death in early spring 2006 was the president of the Kehila Kedosha Janina. Well into his eighties, Genee ran

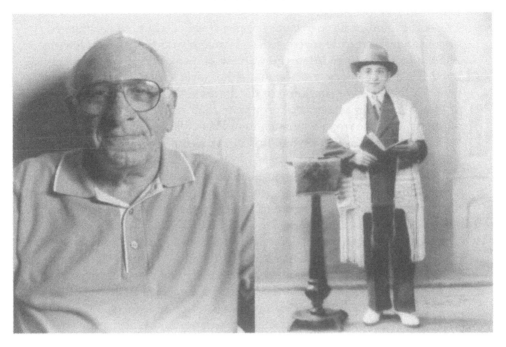

Fig. 1.1. Left, Hyman Genee, 2001, New York City; right, Hyman Genee at his Bar Mitzvah, 1935, New York City. Permission Kehila Kedosha Janina, New York.

the synagogue, served as the reader, and generally kept the congregation in line—it no longer has a rabbi (the community's last rabbi passed away in 2000), and conducts its own services. "You have to make the effort to be here on time," he would reprimand the congregants. Punctuality is not simply a matter of etiquette but also necessity. Nowadays, it can be difficult to gather the ten-man minyan, or quorum, required for services to begin.[9] Those who've arrived on time can sit around for hours waiting for the latecomers before they're able to start. Sometimes extreme measures are taken; if nine men have arrived, and a tenth doesn't seem forthcoming, the congregation may open the curtain covering the Torah scrolls (*heikhal*) and let the scrolls serve as the tenth man. One morning as the minutes ticked by, Hy joked that he was going to go grab someone off the street, "circumcise him on the spot," and have him complete the minyan.

In keeping with patterns on the Lower East Side as a whole, as its Greek Jewish immigrants became established in the United States and accumulated wealth, they moved from Manhattan to the outlying boroughs, New Jersey, and Long Island.[10] After the boom of the 1920s and 1930s, and slight growth as Holocaust survivors came in the late 1940s and early

1950s, the number of congregants began a steady decline, while affiliated organizations that drew together the increasingly far-flung Greek Jewish population in the United States grew. Now known as the United Brotherhood/Good Hope Society of Janina, it has over eleven hundred members.[11] It is the key organization in the States via which the descendants of Greek Jews can express and propagate their identity. The Sisterhood of Janina, Inc., the women's branch founded in 1932, is active in charitable work and has over three hundred members from across the country. The Kehila's actual congregation, however, has all but disappeared.

Today, 280 Broome Street is an unassuming building sandwiched between Chinese food warehouses; the formerly Jewish neighborhood is now overwhelmingly Cantonese. Recently renovated, the synagogue's clean rusty pink facade stands out on a block of dingy tenements and storefronts. The space inside is long and narrow, running north-south, configured with the central reader's platform (bimah) facing north toward the heikhal. (Romaniote synagogues are usually laid east-west, with the heikhal on the eastern wall and the bimah on the western one.)[12] But the Janina congregation's tiny lot size and its orientation led to the Kehila's unusual layout, which is more in keeping with traditional Sephardic synagogues. Upstairs is the women's section (ezrat nashim); the women who sit there today regularly shout down at the reader to speak up, or to add pieces of news they'd like included in the week's announcements. The news reports are not, for the most part, upbeat: word of elderly former congregants who have died or suffered illness, and news of perennial vandalism of Jewish sites in Jannina, back in Greece, feature prominently. Downstairs in the basement is a modest kitchen and a common area where a blessing of bread and wine (kiddush) is held after services. Until the recent renovations, the room's walls were plastered with tourist posters of Greece and Israel; kiddush regularly includes grape juice, shots of ouzo, spanakopita (traditional Greek spinach pie), kalamata olives, hard-boiled eggs, and feta cheese. Bagels or bialys with cream cheese and salmon are a frequent supplement.

In decor and gustatory tastes, as in their self-identification, the Kehila's congregants are at once fully Greek and fully Jewish, and their immigrant culture is interwoven with both Greek and Jewish organizations. The synagogue has sponsored Greek language classes, trips to Greece, and *kafeneíon* evenings designed to replicate the social ambiance of the traditional Greek coffeehouse. Tributes included in the Kehila's seventy-fifth-anniversary commemorative volume are full of Greek turns of phrase: "*Yássou* [Hello] to all" and "*s'agapó*" (I love you); a beloved deceased relative is remembered with the Greek term for grandfather as "a great *Papoo*."[13] As is the case for other diaspora émigré groups, the Greeks of the Janina synagogue are particularly pointed about the expression of their national origins. Some, like Genee, were born in the United States, to parents who came to the States before Jannina became part of Greece in 1913. Others were born in Greece, survived the Holocaust in hiding, and came to the United States in the 1950s and 1960s.

Ilias Hadjis, for example, was born in Athens in 1937, and spent the occupation hiding in the Pilion Mountains and Athens. Koula Cohen Kofinas, born in Larissa in 1938, spent the war posing as a Christian. "I used to go to church and light a candle. First because we didn't have a place to thank God. And second, we didn't want to give [those around us] any reason to think of us as different."[14] Almost twenty years later, she came to the United States. Her husband, Sol Kofinas, who now directs the museum run out of the Broome Street synagogue, spent the occupation hiding in a woodshed with his brother. His father was among those infamously locked in the Athens Melidoni Street synagogue by the Nazis just before Passover in 1944; his mother had fled with the two boys, but was caught by German guards when she sneaked back to the family's house to get some diapers and clothing.[15] Kofinas left Greece in the 1950s. The congregation also includes many Sephardim of Salonikan origin (there was a merger with a Sephardic synagogue in the 1970s), but resolutely sticks to the Romaniote liturgy and insists at all times that it is Greek. When a would-be congregant phoned the synagogue and asked, "You're not Ashkenazic, right? You're Sephardic?" he received the somewhat brusque reply, "No, we're Romaniote."[16]

The congregation's sense of Greekness is fostered by its diasporic, immigrant identity and U.S. context. While in Greece, Jews like Kofinas had a complicated national identity—there was in many settings a vague sense of "us" (Greeks) and "them" (Jews).[17] "Greek Jew" was not an established category. In the United States, and particularly New York City, the Greek immigrant as a category is well-known. The religiously pluralistic surrounding environment doesn't necessarily expect Greeks to be of one religion or another, and Greek Jews in this country more easily inhabit both "halves" of their identity than they did in Greece. Here, their Greekness is not challenged but rather taken as a matter of fact. Meanwhile, Greek Jews back in Greece still struggle today to assert their legitimacy as Greeks. They are often suspected of having divided loyalties between Greece and Israel, or are told that Greekness and Orthodox Christianity are coterminous. The New York context, however, which is far more used to Jews (and far less critical of Israel), more easily embraces the idea of multiple, overlapping identities, such that the members of the Kehila Kedosha Janina are woven into the fabric of an immigrant narrative that has fostered Greek Jewish identity for more than a century.

GREEK JEWS

Hy and his fellow congregants, Romaniote and Sephardic, while now unified by the smallness of their congregation, their shared status as first-, second-, and third-generation immigrants, and their adoption of a unifying national term—Greek Jews—by which to identify themselves, are actually of astonishingly diverse origin. The congregants' distant roots are in a swath of territory reaching from Portugal and Spain, to the northern Balkans, to Alexandria, Egypt. Until the twentieth century, there was no such thing as a Greek Jew. Nor was there a definitive Greece; the country's boundaries were not fixed until after the Second World War. After the Greek War of Independence (1821–33), Greece consisted of small portions of the mainland (see figure 3.1, chapter 3). Over the course of the nineteenth century it grew until, by the end of World War I, it had doubled in size. Finally, after World War II the Dodecanese Islands, of which Rhodes is the largest, were ceded to Greece. Only retroactively have the Jews from these places come to be regarded as a nationalized collective, as Greek Jews. Indeed, Rhodes is a case in point; while at the time of the Nazi deportations from the island Rhodes was actually Italian, its destroyed Jewish community is now classified by Yad Vashem (Israel's official Holocaust "Remembrance Authority") as part of Greek Jewry— an ex post facto classification based on the fact that Rhodes today is part of Greece, and one that says little, if anything, about its Jewish community's own identity.[18]

This awkward category and the communities to which it refers—some of which considered themselves Greek, and some of which did not—reflects the complicated path minority groups negotiated in the transition from a world of empires to one of nation-states. The Jews who today are collectively called Greek—in Israel, the United States, and other places in the diaspora as well as in Greece—are described by a term that would have been meaningless to their ancestors. With the exception of some (largely Romaniote) communities (notably, in Athens and Jannina) in the last decades of the nineteenth century through World War II, the designator would have been met with puzzlement. Salonika, home to sixty thousand Jews before the Holocaust, became part of Greece only in 1912. The Judeo-Greek culture that began to emerge there in the 1920s and 1930s was arguably as much a reflection of assimilation as it was of a new, distinctly Jewish, Greek identity, and was in any case tragically short-lived. Jews from Rhodes were *Rodeslis*, a Turkish word that simply means "from Rhodes." While Rhodes is now included in Yad Vashem's list of destroyed Greek communities, it was not part of Greece until 1947, by which time almost all of its Jews had either left or been murdered by

Fig. 1.2. Hacham Ezra, rabbi of the Kehila Kedosha Janina, 1880s–1925, New York. Permission Kehila Kedosha Janina, New York.

the Nazis. Many Jews from the Island of Corfu—which joined Greece in 1864—historically called themselves Italian or Apulian; those who called themselves Greek (*Griega*) did so in reference only to the language they spoke (and in a typical early modern Mediterranean cultural mélange, used an Italian term to designate their Greekness).

Greek Jews, a unified, nationalized category, emerged out of multiple and fragmented communities, and fully gained purchase, paradoxically, only once more than 90 percent of all Jews actually living in Greece had ceased to exist. The first topoi in which a distinct Greek Jewish identity fully emerged were, like Broome Street's Kehila Kedosha Janina, outside of Greece. It was in Auschwitz and afterward in Palestine/Israel—settings characterized, like New York's Lower East Side, by mass heterogenous Jewish populations—that Jews from Greece came to be defined as a nationalized group, as Greek Jews. One of this book's primary interests is in this phenomenon of what might best be termed "extranational nationalization"—the striking dynamic by which a formerly imperial group of peoples were converted, largely in settings beyond Greece's borders, into a single nationalized people. It is a process that was driven initially by the transition from the Ottoman Empire to the Greek nation-state, but was completed with the Holocaust and the foundation of the state of Israel.

ROMANIOTES AND SEPHARDIM

Roughly speaking, Greek Jews fall into two broad categories: Romaniotes and Sephardim. The older but by far smaller are the Romaniotes, or *Niótes*, as some colloquially refer to themselves.[19] Many Romaniotes believe that their origins are in the Jewish migrations from Palestine that came in the wake of the destruction of the second Jerusalem temple by the Romans in A.D. 70. Some Romaniote communities, most notably in Jannina, in the northwest province Epirus, claim that their ancestors arrived sometime between the Babylonian conquest of Jerusalem (586 B.C.E.) and the early Roman period. Didimoticho, Serres, Salonika, Veroia, Jannina, the Peloponnese, Halkis, and many of the Greek islands were (and in some cases still are) home to Romaniote communities, all of which—despite their communal origin myths—are properly regarded as indigenous to the region.

Romaniote Jews are literally "Roman Jews." Just as most Greek-speaking Orthodox Christians referred to themselves as late as the early nineteenth century as *Romiós* ("Roman"), in reference to their Roman (i.e., early Byzantine) origins, so did Jews of the region. The map of southeastern Europe is dotted with places named "Rumeli" and variants for the same reason; Romania is the most obvious example. In their very name, the Romaniote Jews show their cultural commonality and shared historical origins with a variety of peoples, across a large swath of territory, who trace their origins in Greek lands to the first centuries of the Common Era. The term Romaniote, then, is fairly generic. In the Greek Jewish context, it has evolved in meaning since the sixteenth century to designate non-Sephardic, "native" Jews.

Romaniote Jews were indigenous, in contrast to the Sephardim, whose name—like the Romaniotes'—denotes their place and culture of origin: *Sefarad*, the Iberian Peninsula. Following the Catholic expulsions of Jews from Spain, Portugal, France, and Italy in the late fifteenth and early sixteenth centuries, waves of Sephardim fled east to Ottoman territories. Sultan Bayezit II (1481–1512), eager to restore economic and social balance to the empire in the wake of the successful but draining conquests of his predecessor, Mehmet II, who had consolidated Ottoman rule over the Balkan Peninsula, issued a decree welcoming the Sephardim to Ottoman lands. There they overcame the Romaniotes, emerging as culturally and economically dominant by the end of the seventeenth century. As early as 1509 a Salonikan Sephardic rabbi boasted, "It is well-known that Sephardic Jews . . . in this kingdom . . . [now] comprise the majority here, may the Lord be praised."[20] A few Romaniote communities retained their distinctive "Roman" character, but none was unaffected by the arrival of the Sephardim.

Salonika became the world capital of Ladino rabbinic literary culture.[21] Ladino was its first language until the city underwent rapid Hellenization in the 1920s. For centuries the city was the region's economic center. Sephardic Jews settled in other Greek regions as well. Rhodes, the largest of the Dodecanese islands off Greece's southeastern coast, was a Sephardic center, and small pockets of Sephardim, most engaged in trade, settled throughout the region that today comprises Greece. Many formerly Romaniote communities were replaced with or absorbed by Sephardim.

Smaller numbers of Ashkenazim—central and east European Jews—had also lived on the Balkan Peninsula since the 1400s, when they were ejected from the Kingdom of Bavaria.[22] Cretan rabbinic documents of the early sixteenth century indicate that there were Ashkenazim on the island at that time.[23] Hungarian-speaking Jews settled in Kavála around the same period. Smaller groups included Sicilian and Italian Jews—also referred to in Cretan documents of the early modern era—as well as Jews from Provence and other southern French regions.[24] At first they followed their own liturgy and customs, but like the Romaniotes, were generally gradually assimilated into Sephardic practice as Iberian Jews established their cultural and economy hegemony over the region.

Greek Jews are not a monolithic group, but rather represent a set of cultures that has been a long-standing presence within Greek lands—in some cases, since before their Christianization.[25] In all places but Salonika, Jewish communities were a minority, and usually a tiny one. Some lived in a largely Catholic environment, as on Corfu. Others, as on Venetian Crete, negotiated a Christian environment divided between Catholic and Orthodox Christians, and Muslims. In the Peloponnese, Jews lived in a uniformly Orthodox environment. Some Jews spoke Greek; others used Italian or Ladino. In urban areas, Jewish communities were an integral and central part of economic and mercantile life. In other places, such as early nineteenth-century Athens, only a handful of Jews could be found—largely itinerant traders who temporarily set up business far from home.

In all cases, the size, quality, and vibrancy of Jewish communal life were shaped by political events to both the west and the east, and particularly by the waves of immigration they provoked. One of the characteristic features of Balkan culture generally, arguably experienced most acutely by its Jewish populations, is the profound influence of its location between east and west.[26] From 1500 on, large numbers of Jews from Portugal, Spain, Italy, and southern France made their way eastward to Ottoman territory. Ashkenazic Jews, too, in smaller number, found their way east and south in the wake of pogroms. At the same time, the Ottoman Empire underwent rapid westward expansion. Jewish communities exchanged Catholic and Orthodox rulers for Muslim ones. For the most part, Jewish communities benefited from this change. Centuries later, they were caught between the crosscurrents of western European nationalism and Ottoman "decline." Over centuries Jewish communities, along with other local cultures, were shaped by converging developments to the east and the west.

WHY STUDY GREEK JEWS?

Until recent decades, little scholarly attention was paid to Greek Jews. Jewish studies has long focused largely on Ashkenazim, while Greek histories typically don't include a treatment of Jews—although in recent years this has begun to change, with a number of superb Greek books coming out on specific dimensions of Greek Jewish life as well as important collections of Greek Jewish testimonies. Survey literature has been sparse, with the exception of an excellent French survey of Jewish-Orthodox relations in the modern period, Bernard Pierron's *Juifs et chrétiens de la Grèce moderne*, published in 1996, and shorter essay-style surveys such as Nicholas Stavroulakis's *The Jews of Greece*, first published in 1990.[27] Very little indeed has been written on contemporary Greek Jewry; both Pierron and Stavroulakis stop with the arrival of the

Germans in the 1940s, as do most other works—for a tragically obvious reason. Within Greece itself, this book goes only a bit further in time— to the late 1940s and 1950s, which saw a new host of problems confront Greek Jews, and when a large percentage of survivors emigrated to Palestine/Israel. I make only passing reference to today's Greek Jews, who, between five and six thousand in number, have labored hard to re-create a meaningful communal life in nine of Greece's traditional Jewish communities.[28] Present-day Greek Jewish life is worthy of a scholarly treatment of its own. In both spatial and conceptual scope, however, this book attempts to be more comprehensive than previous works, following Greek Jews as they were deported to Auschwitz and other camps, and as they emigrated to the United States and Palestine/Israel, tracing the development of the Greek Jew as a category in these diasporic settings.

Since the 1980s, Sephardic history generally has received growing attention, mainly because of developments in Jewish studies and Ottoman social and economic history. For the most part, though, these studies look at Balkan Jewish communities largely in isolation, without acknowledging the multicultural and multireligious environment in which they lived—much less its influence on them. This increased interest in the region's Jews also hasn't spilled over to the Romaniotes, whose tiny numbers, cultural obscurity, and eclipse by the Sephardim during the early modern period have often kept them beneath the radar of scholarly investigation.

Interest in Ottoman "multiculturalism," and specifically in the city of Salonika, has recently become intense. It stems in part from a pervasive academic nostalgia for a time when people ostensibly thought differently about religious difference. The collapse of Yugoslavia, the ongoing conflict between Palestine and Israel, and more recently, the simplistic depiction of the world as divided between a "civilized" Judeo-Christian West and "dangerous" Muslim East have fostered interest in a historical moment when Jews, Christians, and Muslims cohabited (it is imagined) with relatively little conflict.

This vision, of course, is romanticized, and downplays the extent to which religious hierarchy and prejudice colored social conditions. But the efflorescence of interest in the late imperial multicultural moment is an important development in the context of a scholarly tradition that inadequately addresses "minority" populations, and in a Western cultural environment that smugly imagines itself to be the first historical instance ever of diversity and tolerance, and too easily glosses over its own prejudices. It is a self-congratulatory tendency to think of multiculturalism as a distinctly modern invention. Yet many of the conflicts routinely termed "age-old" are, in fact, quite young. That groups of people should be categorized and defined along religious lines is an ancient concept; that specific chunks of land and entire nations should be is largely a much newer one. The emergence of the Greek Jew as a category provides one example of the anomalous results of this rapid and dramatic transformation.

In deciding on a title for this book, it was difficult to encapsulate the complexity of this category, or avoid eliding that complexity altogether. It was impossible to avoid teleology. Greece, after all, has only existed for a few short centuries, and Greek Jews, even on a strictly cultural level, haven't existed as such for much longer than a hundred years. This book's primary emphasis is on the modern period, although I attempt to provide a partial history of several of the south Balkan Jewish communities that ultimately made up modernity's Greek Jews, setting these histories in the broader context of Mediterranean history, Ottoman rise and decline, the rise of nationalism, the growth of Zionism, and the internationalization and secularization of Judaism. Its primary backdrop is the hundred-year period during which the Ottoman Empire and Greece existed side by side, during which the region's Jews and Christians alike felt themselves to belong to polities that spanned the two.

The uneven, drawn-out transition from empire to nation, and Greece's century-long period of territorial expansion, resulted in a belated nationalization of Greece's Jews—a nationalization that became complete only with their departure from Greece. It is not by accident that fully half of the history recounted in this book is set outside of Greece: in the concentration camps of the Nazis, and the Greek Jewish diasporas of Israel and New York's Lower East Side. In these settings, communities that for centuries developed identities that varied in their embrace or rejection of Greekness were cast as clearly and uncomplicatedly Greek. This book seeks to show how this is so, tracing the many strands that lead to modernity's paradoxical and largely diasporic Greek Jew, and such groups as the Kehila Kedosha Janina today.

The result has turned out to be a history of Greece, or Greekness, from a different angle. This book presents a history of a group of peoples, but even more, it strives to present a "Jewish history" of both a place and an ideal: Greece. The idea of a Jewish history of Greece will doubtless be an annoying one in some quarters—both those that define Jewish history in academically territorial terms and those that define Greece as something decidedly other than Jewish. But the history of the Ottoman borderlands—like those of various Jewish groups themselves—is nothing if not the history of overlapping categories, national threads, and peoples. It is a history that will be better understood if we are willing to broaden our definitions.

PART ONE

Independence and Expansion

AFTER INDEPENDENCE: "OLD GREECE"

W HEN THE G REEK W AR of Independence broke out in 1821, it was greeted by Western Europe and the United States as a noble battle for democratic freedom, signifying nothing short of the rebirth of the ancient Greeks so celebrated in the Western cultural tradition. As President James Monroe announced to Congress late in 1822, "The name of Greece fills the mind and heart with the highest and noblest sentiments . . . the reappearance of this people in its original character, fighting for its liberty . . . arouse[s] enthusiasm and sympathy everywhere in the United States."[1] One of the realities of the Greek War of Independence, however, was a series of bloody religious massacres. Muslims, Christians, and Jews who had lived on shared terrain for centuries set on one another with astonishing ferocity.[2]

But while Christians and Muslims were clearly affiliated with one side or the other—the Greeks or Ottomans, respectively—Jews, the smallest group by far, were caught in between. At first, they put their money on what they thought would be the winning horse. Assuming that the Greek uprising would fail, when the war erupted most Jews sided with the Ottomans. Historically, Jews of the region had suffered far more from Greek Orthodox Christians than they had from Ottoman Muslims. In theory, if not always in practice, the Ottoman legal system afforded Jews and Christians alike broad freedoms and a relatively high degree of autonomy. Compared to the status of Jews in Christian lands, that of Ottoman Jews was relatively secure, and they feared the prospect of Christian dominion far more than the reality of Ottoman control. At the same time, the Ottomans seemed by far the stronger military power. The decision to side with the Ottomans made sense from every point of view. Yet this course of action quickly proved to be devastating: as the Greek cause gathered steam, Jews were killed as traitors or enemies. In the Peloponnese, where the uprising against the Ottomans first broke out, Jewish residents were slaughtered by klephtic (brigand) fighters and in a series of mini pogroms that swept the region. A long-standing anti-Semitic sentiment was compounded by the Greek anger at the Jewish collusion with the Turks. At the same time, rumors and reports of Jewish collaboration with the Ottomans in other parts of the empire spread to the Peloponnese, provoking further reprisals.

Over the course of the nineteenth century, the pro-Ottoman role—real and imagined—of Jews in the Greek War of Independence was to be invoked by Greek Christians on numerous occasions, held up as supposed evidence of Jewish disloyalty to Greece. The vocabulary of Jewish disloyalty would become a virtual trope of the Greek Jewish experience, recurring during the various Greek territorial expansions of the nineteenth and early twentieth centuries, which brought tens of thousands of Jews into the Greek state. With each expansion—into Thessaly in 1881, and Epirus, Macedonia, and Thrace in 1912—the region's Jews were caught between loyalty to the flagging Ottoman Empire and the realpolitik of the new Greek state.

The War of Independence bore ominous implications for the region's Jews from the start. A week after the revolt was declared, on Easter Sunday, April 22, 1821, the Greek Orthodox patriarch and several bishops were executed in Constantinople by the Ottoman authorities. In an effort to establish the church's loyalty to the Ottomans, the patriarch, Gregory V, had issued a decree excommunicating Greek rebels.[3] Nevertheless, the sultan (Mahmud II) ordered Gregory's execution as a warning to revolutionaries and punishment for the church's failure to prevent the Greek uprising. By widespread accounts, the bodies were put on public display and then turned over to the Jewish community for disposal. Accounts differ as to the precise role played by Jews: some held that a small group was unwillingly forced into participation by the Ottoman authorities; others claimed that out of an eagerness to avenge themselves on Christianity, they had volunteered. The patriarch and prelates' bodies were dragged through the streets, and then thrown into the sea.[4] The story spread throughout the Greek heartland, with the implication that "the Jews"— a collective body—had dealt an unforgivable insult to "the Christians." Only death could avenge it. By the time the news spread west, to the Peloponnese, the role played by Jews in the events had narratively eclipsed even that of the Ottomans. Their slaughter continued apace, justified as retribution for the desecration of the prelates' corpses.[5] Meanwhile, in the northern mainland, Jews, "armed and thirsty for Christian blood," participated in Turkish reprisals against Greeks in Naoussa and shared in plunder taken from decimated Greek villages.[6] This led to further anti-Jewish reprisals.

In the first months of the war, the population of major Peloponnesian towns exploded as Turkish troops and refugees from the countryside poured in. In Tripolis, for instance, in the central Peloponnese, the population virtually doubled between March and November 1821.[7] A number of the refugees were Jews seeking shelter with Tripolis' ancient Romaniote Jewish community. On November 1, Greek revolutionaries laid siege to the city, and its entire Jewish population was eradicated. Some Jews were

armed, but most were not. The famous Greek general Theodoros Koloko-
tronis was particularly contemptuous of Jews who dared to fight; when
he found an armed Jew among a group of surrendering Turks, the Greek
general was outraged. "Bah! An armed Jew, that's not right!"[8]

The destruction of Tripolis' Jews was not simply the collateral damage
of war. The British consul stationed in Patras, in the north, reported the
deliberate massacre of hundreds of Jews; the British consul in Constantinople sent word that thousands of Jews had been tortured and immolated
by the Greek forces.[9] A Corfiote Jewish eyewitness claimed that five thousand Jews were killed in Tripolis, despite the fact that they were noncom-
batants.[10] While figures vary widely, all accounts concur that November
1821 saw the definitive death of a Jewish presence there.[11] "Certain it is,
that the Jewish population of that city, whatever may have been its
amount, was destroyed." As the war spread, Jews around the region suffered a similar end: "The sons of Isaac, and the sons of Ishmael, . . . as on
every occasion during the Greek Revolution, met with a common fate."[12]
Those who survived fled to established Jewish centers, such as Corfu,
Salonika, Istanbul, and Izmir. Some went as far away as Sidon, Tiberias,
and Jerusalem, and others to the Ionian Islands in the northwest.[13] The
Jewish community of Volos, on the east-central coast, was made up entirely of Peloponnesian refugees.[14] At the close of the War of Independence, few Jews remained within Greece's borders. Its largest community,
and one of the few to remain stable in size during the war, was in Halkis,
home to about fifty Romaniote families.[15] No Jews were left in the Peloponnese.[16] In the 1830s, Jewish refugees from the new Greece continued
to move west, to the Ionian Islands, under the British Protectorate, and
north into Thessaly, Epirus, and Macedonia—all regions that remained,
for the time being, under Ottoman control. The very birth of the Greek
state, then, was accompanied by the destruction or flight of the region's
Jewish communities.

As established in 1833, the Kingdom of Greece consisted of Attica, the
Peloponnese, and the Cycladic island group. While small numbers of Jews
were scattered across this territory, all told they numbered fewer than a
thousand, out of a total population of about 750,000.[17] The demographic
reality was in line with the ideological platform of the new Greece, which
as founded in the wake of the Greek War of Independence was to be a
fundamentally Christian nation. The overlay of Greekness with Orthodox
Christianity, enshrined in law in the mid-nineteenth century, when Greek
Orthodoxy was formalized as the state religion, existed in practice from
the start. The Greek War of Independence, despite the fact that many of
its intellectual leaders took the French Revolution as a model, was at the
popular level widely regarded as a fight for Christian freedom from the
"yoke" of Islam.

It is anachronistically modern to dismiss this as "racist" or chauvinist. After all, in the early nineteenth century, the equation of Greekness with Orthodoxy was not simply plausible; any other definition of Greekness was almost inconceivable. The Ottomans had for centuries defined the Greeks according to religion, granting the Greek Orthodox patriarchate virtual autonomy in all dealings internal to the Greek Orthodox community. Law and education—indeed, all matters pertaining to daily civil life—were mediated by the church. Similarly, Jewish life across the empire was mediated by the rabbinate. Imperial rule provided no precedents for a form of national identity that cut across religious lines. Just as the legacy of the millet system left no space for such a thing as a "Greek Muslim," there was not, in the first century of the Greek state, any such thing as a Greek Jew. There were simply Ottoman Jews of various backgrounds who had yet to negotiate a space for themselves in the new nationalized space within which they found themselves.

EXTERNAL PRESSURES

In the uncertain and anxious environment of nineteenth-century Greece, a variety of cultural and political conflicts conspired to make the lives of groups viewed as alien particularly fraught. Early advocates for Greek independence were strongly influenced by the French Enlightenment; Rigas Velestinlis (ca. 1757–98), for example, wrote extensively on human rights, planning a constitution for Greece that was based on the *Déclaration des droits de l'homme*. But later Greek politics was more interested in asserting the primacy of Orthodox Christianity. Ideologues like Markos Renieris (1815–97) argued for the superiority of Christianity over eighteenth-century secularism.[18] While Greece's first proposed constitution (the Epidavros Constitution of 1822, never implemented) reflected Western liberal ideals in specifically calling for the toleration and freedom of all religions, the first constitution actually enacted, reluctantly granted by King Otto in 1844 after a protest coup staged the year before, declared Orthodoxy the official state religion and forbade other religions the right to engage in proselytizing activity. For the next four decades, Jews in Greece had no formal recognition whatsoever from the state.

The new nation's first decade (1833–43) was characterized by intense economic, political, and cultural pressures. Greece was virtually bankrupt, subject to crushing Western domination, and bereft of viable industry. All major Greek trade centers—Salonika/Selânik, Constantinople/Istanbul, Alexandria/Iskenderiye, and Smyrna/Izmir—remained in Ottoman territory. The half-starved population, disproportionately comprising women and children, was largely landless and illiterate. Fully 75 percent of the world's Greek-speaking population lived outside the new nation's borders.[19] The tiny country itself was home to fewer than one million citizens. Economically, the country was scarcely viable. The international pressures were also tremendous: Greece existed under the "guarantee" of the three "protecting powers," Great Britain, France, and Russia, which together set Greece's boundaries.[20] The Conference of London (May 7, 1832) had declared a foreigner, Otto, the adolescent second son of Bavaria's King Ludwig, king of the Greeks.[21] Each of the great powers had tried to establish hegemony in Greece. The Russians, for example, had for a century gone back and forth between behaving as Greece's ally and as part of a unified bloc, along with France and Britain, that aimed to control Greece's domestic affairs. In 1770, they had urged the Greeks to take up arms against the Ottoman Empire, appealing to their shared status as Orthodox Christians and making false claims of military support in the uprising. Again during the Greek Revolution (1821–33), the Russians cast themselves as the great supporters of the Greek people, and after the conclusion of the war had attempted to cultivate and appeal to a Greek sense of indebtedness. So too had the British and the French. This had led to a most peculiar domestic political situation in Greece: the Greek government was identified as the "Russian" party, while the opposition party, under the leadership of Ioannis Kolettis, came to be known as the "French" party, in part because of the close diplomatic ties Kolettis had enjoyed with France in the final years of Ottoman rule when he served in the court of Ali Pasha of Jannina. Because of this state of affairs, the British, under Lord Palmerston, unhappily pushed for the development of another opposition party, this under the leadership of Alexandros Mavrokordatos, which came—to Palmerston's chagrin—to be known as the "English" party.

External influence rested on Greece's immense monetary indebtedness to the great powers. Otto's kingship was conditioned on a massive loan, compounding the vast international debt Greece had racked up even before the war was concluded.[22] The Greeks were inexperienced with the European environment, which was flush with surplus cash from its recent colonial successes. Loans were negotiated on staggeringly bad terms for the Greeks, with the Greeks effectively signing away everything they had before their nation had even come into existence. As printed on the bonds issued during the War of Independence, the Greek guarantee of payment read: "To the payment of the annuities are appropriated all the revenues of Greece. The whole of the national property of Greece is hereby pledged to the holders of all obligations granted in virtue of this loan, until the whole of the capital which such obligations represent shall be discharged."[23] Greek "independence" meant freedom from Ottoman control, but neither political nor economic autonomy. As the British minister to Greece commented, "A Greece truly independent is an absurdity; Greece is Russian or she is English; and since she must not be Russian, it is necessary that she be English."[24]

GREEK CHURCH AND STATE

These pressures conspired to foster intense Greek resentment at what was widely perceived as an attack on Greek cultural autonomy—a view fostered by the fact that the great powers had tried hard to keep the power of the Greek Orthodox Church at a minimum. While under the Ottomans the patriarchate had held authority over all Orthodox Christians, the new autocephalous Church of Greece, established in 1833, was responsible only for Greek citizens, and was subordinate to the government. Far more Orthodox Christians lived beyond the borders of Greece than within them. Greece's new leaders wanted to put an end to the centrality of the church in political affairs, and to effect separation from the Constantinopolitan patriarchate. The Catholic Bavarians, along with many Western philhellene observers, regarded Greek Orthodoxy with contempt, viewing it as a mixture of ignorant pagan superstition and incorrect Christian doctrine.

An act of April 4, 1833, severed the Greek church's ties to the patriarchate and placed it under the control of the Greek state. This development was taken as clear evidence of the dangers and depravity of Western liberal notions. It also sent a signal to both the patriarchate in Constantinople and Russia that Orthodoxy was no longer to be invoked as the pretext for external intervention in Greek affairs. Henceforth, the church had as its mandate simply to minister to the spiritual needs of the nation, and was to be one among many state-administered institutions. Clerics of the Holy Synod would be responsible for church governance along with liturgical and dogmatic decisions, but have no political power. Not since the conversion of Constantine had the church been divorced from a political role. Throughout Ottoman rule the patriarch, an appointee of the sultan, had served as political as well as spiritual leader of the Orthodox people. For Greek church hierarchs, the new arrangement was almost intolerable.

It was also unpopular with the general populace and political conservatives, who regarded the church as a bulwark against the modernizing influences of Western thought. Monks, in particular, were bitterly opposed to the church settlement of 1833—the number of monasteries had been reduced by two-thirds by the government—and agitated on behalf of Russia in the power struggle between the Russians, British, and French for influence in Greek affairs. They argued that the settlement reflected the evil interventions of "Jews and secularists"—claims that fanned the burning flames of postwar anti-Semitism and capitalized on a growing climate of xenophobia.[25]

Greek Christians also had to contend with the arrival of a slew of determined Protestant missionaries. Mainly British and American, they hoped to cleanse the Greeks of superstition, introduce them to the "true" Gospel, and redeem them from abject poverty. They represented yet another form of external interference. Episcopal missionaries from the United States, for example, took up residence in Athens in 1831, opening a school to train "the rising generation of excitable Greeks."[26] The school had as a primary curricular feature an emphasis on the rightful separation of church and state: "Mrs. Hill wisely endeavored to impress upon [the students'] minds a respect for constitutional authorities, particularly important in that agitated and unsettled community."[27] Such missionaries were seen as both religiously and politically intrusive, fomenting Greeks to leave their native faith and adopt Western liberal ideals. The Church of Greece took measures to prevent the Protestant incursion. In 1836 and again three years later, the Holy Synod issued patriarchal encyclicals condemning the translation of the New Testament into demotic Greek. The original Greek of its composition—*koine*, or "common" Greek, the lingua franca of the late Hellenistic period—was unintelligible to many Greeks, and Western missionaries had tried to capitalize on this by commissioning translations of the Gospels into the modern "demotic" version of the language spoken by most Greeks of the region. The Calvinist and Protestant Foreign Bible Societies, and their translation activity in particular, undermined neo-Hellenism's emphasis on the Greek language and the Orthodox faith as the twin pillars of Greek identity. In response to these multiple pressures placed on it in the wake of the War of Independence, over the nineteenth century the autocephalous church pushed for greater influence through an aggressive program of Hellenization, linking Greekness to Orthodox Christianity more tightly than ever before. The linkage of Orthodoxy and ethnos, along with the virtual nonexistence of Jews in Greece, rendered the notion of Greek Jews a conceptual oxymoron, while demographics guaranteed that for the time being, they were a nonissue. Not until the 1860s would Greek constitutional law recognize them formally. And even then, the shift had to do not so much with debates over the status of Jews per se as with a fundamental sea change in the nature of the Greek state—which under the 1864 constitution became a constitutional "crowned democracy,"[28] where power rested with the people rather than the monarch.

ATHENS

Soon after Greece was established as an independent kingdom, Athens was designated the capital.[29] Here, too, Western influence and philhellenism won the day: while Greeks argued for a capital in Nafplion, on the eastern coast of the Peloponnese, to the great powers it seemed inconceivable that any place but Athens would be Greece's capital. But redolent as it was of the glorious classical past, in the 1830s Athens was little more than a miserable village, ruined by years of war, with no real cultural or economic significance. While it had a Jewish presence as early as the Hellenistic age, in the early nineteenth century no Jews lived there.[30] In the seventeenth century, as a result of Jewish migrations from Spain, some twenty Sephardic families settled in Athens.[31] But by the late eighteenth century any Jews had disappeared as the city, gutted by years of war, sank into cultural and economic obscurity, and its total population dwindled to a few thousand.

The Jews who came to Athens in the first decades of Greek independence were not Greek in any sense of the term: they came from western Europe, did not speak Greek, and were the subjects of other nations. Athens in the middle of the nineteenth century saw booming growth, and various foreigners, Jews among them, filtered in to take advantage of the economic possibilities presented by the rapid development. Typical of them was Max Rothschild, a member of the king's Bavarian entourage and from one of the most prominent Jewish families in Europe. Rothschild was the first settler in Athens' fledgling new Jewish community, and at Otto's request became president of the Athenian Jewish community—a position that remained informal in the eyes of the Greek government for most of the nineteenth century; the state formally recognized the community only in 1894, by which time Jewish immigration from Jannina and Asia Minor, among other places, had swelled its ranks to close to five hundred.[32] To the degree that he was aware of them, Otto was relatively sympathetic to the Jews of his kingdom, and the few Romaniotes left scattered around the core territories that made up the initial Greek state likely found the postrevolutionary period a welcome change from the turmoil and massacres of the revolution.[33]

The first Jews to arrive in Athens were Ashkenazim, and the first community was made up of foreign Jewish businesspeople. Most were representatives of important central European trading families; some ultimately became members of Athens' informal "aristocracy" because of their huge wealth. On top of the Ashkenazic presence was laid a Sephardic one from other Greek regions, notably Izmir/Smyrna. The Sephardim weren't as wealthy as the Ashkenazim, and were less integrated into broader Athenian society. They also came for economic reasons. Neither group had a synagogue or state recognition. Their small numbers, plus the fact that all were newcomers, furthered their obscurity and fostered the development of self-sustaining religious organizations. By 1878, about sixty Jewish families had settled in Athens.[34] Since all were recently arrived immigrants from an array of backgrounds, the intercommunal subgroupings that dominated Salonika were absent.

The ongoing lack of formal recognition and the transient nature of the early community put these first Athenian Jews at a social disadvantage. They couldn't negotiate legal matters corporately and also lacked official mechanisms for receiving charitable assistance from outsiders. When in 1843 the eccentric philo-Semitic duchesse de Plaisance, Sophie de Marbois, wanted to give the community a large tract of land in Zappeion for the construction of a synagogue, the gift was blocked by Greek law.[35] Prominent individuals in the community undertook fund-raising efforts on their own to support charity and the construction of communal buildings, and turned to Jewish communities in the outside world, particularly Florence, Paris, Trieste, and Venice, for funds.[36]

The economic and political chaos of the postwar decades, the efforts of the church to reassert its authority, and the consolidation of a Greek national identity that was increasingly Orthodox in character all created a fraught environment. Jewish life consisted of tiny indigenous outposts in the hinterlands and a growing foreign Jewish presence in the capital. There was virtually no communication between the two groups. Athens' community maintained contacts with Western Europe, while the smattering of Jews in the smaller towns of the mainland were linked to Jewish communities in the Ottoman Empire. There was no consolidation of Greek Jewry, either in institutional or cultural terms, and the term Greek Jew would have made little sense to anyone in Greece, Jew and Christian alike.

THE "PACIFICO AFFAIR"

Greece's growing resentment of external meddlings in its affairs, coupled with the consolidation of a Hellenism heavily inflected by Orthodox Christianity, conspired to create a potentially xenophobic environment. Individuals who were perceived as economically superior, allied with the Western powers, or threatening to Orthodox Christianity were viewed with resentment and paranoia. In 1847, a series of events coalesced around a figure who was seen as all three. The story of Don Pacifico, a Jewish British national living in mid-nineteenth-century Athens, gives a vivid example of the complicated matrix within which Jews in the new Greece were embedded.[37] It also shows how difficult it is when considering the period to tease apart anti-Semitism from the economic, political, and demographic strains brought by the rapid consolidation of the Greek nation-state. Finally, Don Pacifico's story reveals the extent to which it was assumed in the period that Greek and Orthodox Christian were coterminous categories.

"Twenty-two soup spoons, silver, of English provenance; twenty-four forks; a sugar pot, with silver tongs; a milk jug; a salt cellar; twelve teaspoons, and salt spoons; a large platter; a bowl; a pair of candlesticks belonging to Madame Pacifico."—thus wrote the "Chevalier" David Pacifico from Athens the week of August 14, 1847.[38] "Jewels, diamonds, pearls of Madame Pacifico and her daughters," he continued, "bedclothes and tablecloths," "porcelains," "bronzes," and "household possessions."[39] All, Pacifico wrote, were destroyed or stolen. In a second document he listed the costs of repairs on his home: sixteen windows, with grilles and shutters; cornices; paint; gesso; new kitchen and bathroom—a complete overhaul and reconstruction.[40]

Pacifico, a Jewish resident of the Greek capital, formerly in the employ of the Portuguese diplomatic corps, was writing to enumerate the losses he had suffered at the hands of "the populace of Athens, last April 4th"—that is, Easter Day, 1847.[41] This collectivity—the populace of Athens—would by the end of the affair be but one invoked in the case of Don Pacifico. Britain, Portugal, Bavaria, Russia, France, "the Israelite community," and Greece would all be enumerated in the subsequent events as perpetrator, accomplice, or victim. The episode was embedded in the complex matrix of international legal and economic concerns that dominated Greece at the time. Amid the storm of it all, Pacifico's Jewishness was a lightning rod that attracted attention from many directions. The Pacifico affair, as it came to be known in British diplomatic circles, provides a microcosm of the fault lines that underlay Greek society over the course of its fiercely nationalizing nineteenth century.

The losses to which Pacifico had been subject, and for which he was to spend the better part of the rest of his life seeking redress, were the result of what today would be called a hate crime. The British consul in Athens, Edmund Lyons, was to characterize it—in an angry letter to Greece's first prime minister, John Kolettis—as "one of the most barbarous outrages ever committed in modern times."[42] But while the most immediate contributing factor in Pacifico's losses was the fact that he was Jewish, the full dimensions of his case were ultimately to reach beyond religious categories, and into international and diplomatic ones. Pacifico, a "transnational" Jew in the new Greek state, had come up against an array of conflicting political and social forces—forces that over the course of the next century would confront almost all minorities living in Greece.

On Easter Day, 1847, an angry crowd gathered outside Pacifico's central Athens home, broke down his door, took what silver and jewelry they could lay their hands on, shredded various documents, and then set fire to the building. Among the participants in the mob violence were a number of Athenian youths from high-profile families, most notably the son of Kitsos Tzavellas, then the Greek minister of war and later successor to Kolettis as prime minister.[43] Pacifico sent immediately for the police, who in due time arrived, but then, according to reports, did nothing to prevent the ongoing riot. Witnesses recount that the police stood by watching for several hours before they made any attempt whatsoever to intervene. As Pacifico himself put it, the crime had been committed "in broad daylight, before the very eyes of the authorities."[44] Indeed, the incident was so prolonged that all the important personages in Athens—from the British consul to the Greek prime minister—learned of it even as it was still under way.[45]

Pacifico's complaint, however, was not made only on his own behalf. Among his losses were 8,266.54 drachma (£269.19.6) worth of silver, contributions entrusted to him by various international Jewish communities.[46] The money, by Pacifico's wording, had been stolen by the populace of Athens. The case, then, was also a sort of transnational class action suit, brought by Pacifico against the city of Athens, on behalf of the international Jewish community. As Pacifico explained, the Greeks had stolen the Jews' money. The case file included copies of letters written by the leaders of various prominent Mediterranean Jewish communities committing funds to assist with the construction of a new synagogue in Athens.[47] Pacifico's victimization was the product of what he called "fanatisme"— fanaticism, a violent attitude and behavior that afflicted, in Pacifico's experience, much of the Athenian elite.[48] The sentiment, he argued, was one commonly directed by Greeks at Jews. And in his view, he alone was not the sole Jewish victim of the crime. So too was the whole "Israelite community" of Athens, along with its international supporters.

Those involved in the assaults on his home—all children of high-ranking Athenian politicos and aristocrats—were well-placed to have heard of Pacifico and know his reputation as a wealthy troublemaker, which stemmed from a variety of suits he had brought against the Athens government and a shady history in Portugal. In the 1840s Athens was home to some twenty thousand people.[49] Its wealthy elites, who included both Pacifico and his assailants, all lived within close range of one another in the rapidly developing neoclassic neighborhoods on the city's eastern limits. The district was relatively cosmopolitan, including consular residences and offices, but everyone within it knew everyone else's business, and gossip about others' affairs was rampant. Pacifico was not, however, responsible for the offense of which the mob felt he was the perpetrator, about which allegation and rumor had spread like wildfire on Easter Day.

In 1847, for the first time in the history of the Greek state, the Greek authorities had issued a ban on the traditional Easter festivities, which had as their centerpiece the burning of an effigy of Judas, and were marked by a weeklong free-for-all assault on Jews by Orthodox Christians.[50] The mob scene at Pacifico's house echoes an array of accounts, from around Greece, of the early modern and modern periods: "A large crowd gathered outside D. Pacifico's home and apparently a few children began throwing stones at the house, and then many men threw themselves against the doors, and, breaking the windows, entered."[51] Several hundred people were involved—a veritable mob.[52] The angry crowd was enraged to have been forbidden the practice of burning the effigy of Judas and held Pacifico responsible. Not realizing the ban on effigy burning was widespread law, the members of the Greek mob thought instead that it had come about through some financial arrangement between Pacifico and the local parish church, that "the Jew Pacifico, by paying off the churchwardens of [the local] church, had succeeded in preventing the burning of the image of Judas, which they did by custom each year."[53] Three years of litigation between Pacifico and the Greek government ensued. The ultimate outcome was Palmerston's 1850 order of an embargo on all Greek vessels and a major international incident. Insofar as the episode is known today, it is in the context of Palmerstonian "gunboat diplomacy." But it also illustrates vividly the complicated social context of midcentury Athens, and particularly the convergence of the decline of the formal status of the Greek church, on the one hand, and the strong Greek Orthodox national sentiment at the popular level, on the other.

What had happened to Pacifico and his family, while horrible, will be familiar to anyone with even a passing knowledge of early modern European history. There are numerous examples from an array of Jewish communities living in Christian contexts of anti-Semitic outbreaks corresponding to the Easter season. In Athens, as in many Orthodox communities, the week before Easter was each year marked by ritual violent outbursts of anti-Jewish sentiment. Culminating in Good Friday, the day of the crucifixion, Holy Week marks the liturgical reenactment of the final days of Jesus's life. In theory a time for somber reflection and lamentation, in practice the emphasis is placed instead on the perpetrators of Jesus's crucifixion. The one individual on whom the greatest anger is poured is Judas, identified in the synoptic gospels simply as "the betrayer."[54] Yet wrath is reserved also for the Jewish people as a whole, portrayed in the Gospel account as the collective murderers of Jesus.[55] Most inflammatory was the Greek Orthodox liturgy for Holy Week services, which explicitly identified "the Jews" as the betrayers of Jesus and held them responsible for the crucifixion.[56] While the region's Orthodox Christians mourned the death and celebrated the resurrection of Jesus, its Jews—quite prudently—would shutter their houses and remain inside in anticipation of the widespread Christian vandalizing of Jewish neighborhoods.

Not until 1890 was there an official, blanket church ruling on the link between these texts and Holy Week anti-Jewish violence; forty-three years after Pacifico's appeal, the Holy Synod issued an encyclical condemning the practice of burning Judas in effigy. In the attempt to discourage it, the synod identified it as a product of Ottoman influence—the one surefire way to make Greeks turn away from something being to paint it with a Turkish brush. In the same encyclical, the synod stated that in accordance with the Gospels, Christians were compelled to "show tolerance towards the Jews, who are the creatures of God."[57] That the two issues—the Judas effigy and the treatment of Jews—were mentioned in the same decree reveals the extent to which the practice of effigy burning functioned as an incitement to anti-Semitism.

Yet the degree to which the event was merely anti-Semitic should be questioned.[58] Its communal dimensions are clear and need not be elaborated. But other factors were also at play. Athens in the mid-nineteenth century was a notoriously violent and crime-ridden place; homes were broken into and sacked regularly. After the city was designated as the national capital in 1834, a decades-long period of construction and immigration began. With the rapid growth, violent crime rates soared.[59] Domestic theft was common. As one visitor observed, "In every house great precautions are adopted against robberies [which recently] were frequent. . . . A band of ten or fifteen robbers has been known to enter one of the largest houses in the city, and to plunder it of all its valuables. The poorer class of houses is entered with comparative ease."[60]

The heavy hand of international intervention in Greek domestic affairs and the quasi-colonial relationship exercised by the great powers over Greece were another element. In late March 1847, the Greek government had been paid a visit by the British lord Lionel de Rothschild, whose family had a decade earlier extended a hefty loan to the Greek government when Britain, France, and Russia had said they would serve as its guarantors.[61] When Greece had defaulted on payments, the British had paid Rothschild on Greece's behalf, and now both Rothschild and Britain began to put heavier pressure on the Greeks.[62] By spring 1847, Rothschild felt that a personal visit to Athens to tend to the matter was a pressing necessity. The Greek ban on the effigy burning was an attempt to avoid insulting Rothschild during a visit dealing with the sensitive matter of finances. But at the same time that the government didn't want to offend Rothschild, it resented him tremendously for the economic hold he and the great powers had over Greece, and for his close ties to the Greek royal family.

Finally, Pacifico himself was a highly unpopular figure in town. He had brought other lawsuits against the Greek government and had tried to sue Portugal as well.[63] He was known as a litigious and opportunistic dissembler, and had misrepresented his credentials and generally earned the dislike of a number of his neighbors. While he himself was eager to present what had happened as stemming solely from his Jewishness, Pacifico had plenty of enemies, Jewish and non-Jewish alike. The fact that he was a Jew was not the only thing that set him apart from his neighbors; he had served as a Portuguese diplomat in Athens and emphasized that his Gibraltarian birth made him a British subject. Pacifico himself explained that his vulnerability derived primarily from the fact that he was "far from my homeland, in a foreign country."[64] In asking for assistance from Britain, although it may well have been needed, Pacifico clearly aimed to capitalize on the fact that Greece was under the British thumb both politically and economically. For their part, the British, too, were clearly using Pacifico as an occasion to exercise the Palmerstonian principle that British subjects, no matter where on the globe they might find themselves, enjoyed the protections of the Crown. To be sure, they presented his case as a matter of "the rights of man" and scorned the Greeks for their ignorance of the principle of "religious tolerance."[65] But in reality, it was one of a series of episodes in which Britain took up the cause of nationals abroad so as to demonstrate its international supremacy.[66] What was a stake for the British was not so much the assertion of religious tolerance as of British primacy.

While the Greek state accused Pacifico of disloyalty to Greece—the land in which, after all, he now lived—for dispensing with its legal system in favor of British diplomacy, at the same time it repeatedly emphasized that it could not interfere in Pacifico's affairs. Both as a foreigner and a Jew, Pacifico did not have the protection of Greek law. But he was foreign in part precisely *because* he was Jewish; as Pacifico wrote, "The fact that [he was] Jewish" meant that he was "not qualified to behave like a Greek citizen."[67] By his reasoning, it was the complete exclusion of Jews from Greek life that made the Greek government the true agent of the crime against him.[68] No Jew, he insinuated, could ever hope to be considered Greek, nor would one wish to be. To an extent, the Pacifico affair can be read as illustrative of the degree to which juridical and social categories in the new Greek state couldn't accommodate the prospect of a Greek Jew.

The assault on Pacifico's home and family was, to be sure, motivated by prejudice and anti-Semitism. The Greek government's unwillingness to press his case rested on the same. But there were many other factors at play as well: anxiety about the diminished role of the Greek church in civic life, huge resentment at the outside political and economic pressures, and the coterminous overlay of Greekness with Orthodoxy. In a setting in which Jews were a minuscule minority with no formal status, legal matters relating to them were settled on an ad hoc and individual basis for the first decades of Greece's existence.

THE "GREAT IDEA" AND TERRITORIAL EXPANSION

The transition to nation from imperial province was not a clean or abrupt one. For close to a century, until the founding of the Republic of Turkey in 1923, Greece and the Ottoman Empire existed side by side. Following the Ottoman reforms of the mid-nineteenth century, which granted citizenship to non-Muslim subjects, Ottoman Greeks enjoyed a sort of dual membership in both the Greek Kingdom and the Ottoman Empire— Greek citizens could not be Ottoman citizens as well, but did have the right (or rather, obligation) to be members of the "Rum millet," the Ottoman Orthodox Christian collective. As such, they were at once Greek citizens and formal "ex subjects" of the Ottoman Empire.[69] For their part, Jews, too, were more transnational than national—that is, national borders were far less important than communal ones. Links between different communities in Greece were not as strong as those with communities in the Ottoman Empire. Jews in Greece maintained rabbinic and trade networks throughout the Mediterranean, as they had done for centuries. The center of gravity for Jewish life remained Ottoman, revolving around such centers as Salonika, Constantinople, and Smyrna.

And just as the empire did not vanish quickly, Greece did not come into existence overnight. The *Definitive Settlement of the Continental Limits of Greece*, as the 1832 agreement termed it, would prove to be anything but.[70] Over the course of the next 115 years, Greece's borders would be redrawn six times.[71] By the close of the Balkan Wars in 1913, Greece's Jewish population would swell from a few hundred to close to one hundred thousand. "Old Greece," as the original Greek territories came to be known, represented less than one-half of the total landmass that ultimately would make up the country. Jewish communities lived throughout the region, and while Greek Christians were happy to see the gradual expansion of the new Greece, many Jews who hoped to maintain the status quo of Ottoman rule were not.

Expansion was driven by the so-called Great Idea (*Megáli Idéa*), a theologically tinged doctrine that aimed for the territorial re-creation of the Byzantine Empire. Tiny, and with few economic and intellectual resources, Greece could ill afford to define as Greeks only those who had actually been born there. In a speech to Parliament in 1844, now much-cited, Kolettis articulated what was to become the basis for Greek policy on this issue: the nation's primary concern was not simply "the fortunes of Greece, but of the Greek race."[72] By this view, Greeks born outside the boundaries of the new Greece ("heterochthons") were Greek nationals no less than those born within Greece's borders ("autochthons"). Many prominent Greeks—Kolettis, the first prime minister, among them—were heterochthons. This transnational formulation served their interests by justifying their increasing hold over Greek elite society.

Heterochthons tended to be wealthier and better educated than autoch-thons, and by the mid-nineteenth century dominated Greek political and intellectual life. In contrast, war veterans were largely poor and illiterate. They tended to be religious, fiercely nationalistic, and to understand Greek liberation not through the liberal democratic vocabulary of the French Revolution but rather as the triumph of Christian good over Mus-lim tyranny. Resentment between heterochthons and autochthons grew as the fighters of the War of Independence were replaced by a new, Euro-pean-trained army, and Greeks from Constantinople and foreigners from the West poured in and took control. As the celebrated war general Ioannis Makriyannis, himself an autochthon, bitterly commented, "As for . . . those who sacrificed [for their country] . . . let them loiter barefoot and wretched in the streets and cry for their bread . . . [while] the filth of Constantinople and Europe abound. . . . They are our masters and we their serfs. . . . They took the finest sites for their houses and took fat salaries in the ministries. . . . All of them became men of property. [But] wherever the [autochthon] Greeks go they suffer."[73]

Greek society was divided between heterochthon and autochthon, Westernizer and traditionalist, the new elites and the old peasantry. These conflicts generated xenophobia and placed an increased emphasis on Or-thodox Christianity as the common denominator that defined all Greeks, heterochthon and autochthon alike. Earlier liberal efforts to sideline Or-thodoxy's influence had already led to a "philorthodox" movement, fed by the perception that Orthodox Christianity was under siege. The move-ment aggressively asserted Orthodoxy's centrality to Greekness, and pres-sured King Otto to set aside Western agendas and bow to the will of the Greek people themselves. The philorthodox movement coalesced around a society of the same name, founded with the stated purpose of liberating the Ottoman provinces of Macedonia, Epirus, and Thessaly. The society's claim on Greece's Orthodox "coreligionists" beyond the kingdom's ac-tual borders converged with the Great Idea's dream of a reconstitution of Byzantium, with Constantinople as its center. The Great Idea also pro-vided a vehicle for bridging the divide between heterochthons and au-thochthons by including all territories ever inhabited by Greeks in an imagined Greater Greece, of which all Orthodox were in theory citizens. When it came to Greeks, such notions were theoretically inclusive. But for others, they created a climate of implied exclusivity. And while the population of the Old Greece was overwhelmingly Greek Orthodox, some of the new territories Greece would grow to acquire were far more diverse. Greek nationalism's increasing reference to Orthodox Christianity would make the transition from Ottoman subjects to Greek citizens easy for Christians, but complex for Jews. Greek Jews would emerge as one of the logical, if taxonomically unprecedented, products of the transition from empire to nation.

Chapter 3

"NEW GREECE": GREEK TERRITORIAL EXPANSION

IN THE DECADES before and after the turn of the nineteenth century Greece doubled in size. Expansion, however, was not a smooth or uniform process. Different territories became Greek in different ways. After centuries of Ottoman rule, Thessaly (1881), Epirus (1913), Macedonia (1913), and Thrace (1920), along with many northern and eastern islands, passed directly from Ottoman to Greek hands during military territorial expansion. The Ionian Islands, in the Adriatic Sea off the west coast, were never Ottoman, going from Venetian, to French, to Russo-Turkish, to British control before being joined to Greece in 1864. The Greek Revolution (1862), combined with a unanimous vote of the Ionian assembly, brought about their ultimate unification with Greece. Crete became Greek (1913) as a result of overwhelming agitation among its population for unification (*énosis*). Rhodes and the Dodecanese became Greek by international postwar settlement in 1947, after becoming an Italian possession in 1912 following four centuries of Ottoman rule.

The transition was greeted in an array of ways by the Jewish populations of these regions. In some, notably around Jannina in Epirus, Jews fought fiercely alongside Greek troops in the effort to shrug off Ottoman control.[1] After Epirus became part of Greece in 1913, Jannina's Jews were formally thanked by the Greek government for their participation on the Greek side of the battle between Greece and the Ottoman Empire; when Crown Prince Constantine entered the town in February 1913, he attended a special service in the synagogue in salute to the community's assistance during the battle.[2] Until the present day, members of the community recall Janniote Jews as having consistently been philhellenes.[3] In other places—Salonika, most notably—Jews were fearful of falling under Greek rule and raised money to help the Ottomans rebuff Greek offensives. When Macedonia was annexed by Greece in 1912, Jews who had taken the Ottoman side in Greek expansionist battles, as in the Peloponnese during the Greek War of Independence ninety years early, were later compromised for having done so. In Thrace, the status of which wasn't finalized until 1923, Jewish communities were able to remain relatively unaffected by incorporation into Greece and retained much of their Ottoman character until their destruction by the Nazis in the Second World War.

Fig. 3.1. The expansion of the Greek state, 1832–1947. From Richard Clogg, *A Concise History of Greece* (Cambridge: Cambridge University Press, 1992), 43. Permission Cambridge University Press.

The circumstances under which different Jewish communities "became Greek" conditioned the transition as relatively traumatic or smooth, as did historical factors. While the period of independence had been universally devastating to Jews living in what became Greek territory, the impact of Greek territorial expansion was far from uniform. Communities, as in Jannina, that spoke Greek were more easily able to imagine life as Greek citizens. Those that had a history of particularly marked tension between Christians and Jews, as was the case in Corfu, tended to be more adversely affected by the transition. Where economic interests were at stake, as in Salonika, the transition from empire to nation-state heightened intracommunal conflict; indeed, in many instances conflicts that were expressed in the language of communal differences were in fact rooted in economic competition.

The process by which the Jews of the "New Greece" became Greek was as uneven as the emergence of Greece itself. Uneven, too, was the extent to which Jews around the country would develop a distinctly Greek national consciousness in the decades after becoming Greek citizens. The heterogeneity of the various processes at play in Jewish integration is perhaps best exemplified by the case of Corfu. As different territories were integrated into the Greek state, complex and variegated Jewish histories and peoples were gradually centralized into *a* Greek Jewish community. While Corfu is the most extreme instance of this process at work, the heterogeneity and complexity of the Corfiote Jewish case was present, to varying degrees, around the country.

INTERCOMMUNAL AND INTRACOMMUNAL CONFLICTS: THE CORFIOTE EXAMPLE

As evidenced in the Greek War of Independence, friction between Greek Orthodox Christians and Jews was commonplace in the south Balkans, although it generally took external pressure to bring it to the level of violence. In regions with a pronounced history of anti-Semitism, the advent of Greek rule was met with trepidation on the part of Jews, who feared a loss of privileges, security, and legal status conferred on them by imperial regimes. Corfu is perhaps the starkest example, with a centuries-long rabbinic record that testifies to the measures taken by the community to defend itself from what it called its Christian "enemies"—the Greeks beside whom the community lived.

Corfu, along with an array of Greek Jewish centers, illustrates the extent to which Jewish fortunes in parts of the South Balkans can be indexed against the gradual expansion of the Greek state. Before unification with Greece, between 1802 and 1863 Corfu's Jewish population rose from 1,229 to close to 6,000 as Jewish refugees fleeing the Greek War of Independence arrived from the Peloponnese.[4] The exact opposite trend came in the wake of Corfu's unification with Greece in 1864; within the next fifty years, the Jewish population fell by close to two-thirds.[5] Other Jewish centers show similar figures. Jannina's Jewish population, for instance, grew by more than a thousand in the wake of Greek independence, and then in the fifteen years following Jannina's unification with Greece (1912), shrank by up to a third.[6] Overall, 30 percent of Epirus' total population emigrated between 1895 and 1906, but among the Jewish population, percentages were far higher.[7] While about 6,000 Jews lived in the city in 1900, by 1930 there were less that 3,000.[8] Salonika's Jewish population rose sharply after 1833, and then began a steady decline after the Greek conquest in 1912.[9] In each case, various factors were at play in the changing population figures, but a constant in each is the impact first of Greek independence and then of Greek expansion.

Greek expansion also closely tracked anti-Semitic outbursts, which seem (briefly) to have proliferated in many places in the wake of unification with Greece. In Volos, annexed to Greece in 1881, a blood libel and series of minipogroms against Jews were perpetrated in 1889 (but seems not to have led people to flee). A similar episode took place in the village of Pogoni, north of Jannina, in 1919.[10] In the case of Corfu, a decisive factor in out-migration was an episode of blood libel in 1891, in which close to twenty Jews were killed.[11] Immediately after, more than fifteen hundred members of the community fled the island in one week alone.[12] While Corfu's Jewish population would increase slightly with Sephardic emigration from Salonika in the late 1920s after the Asia Minor catastrophe, by the start of the German occupation there were only about nineteen hundred Jews living on the island, fully two-thirds less than there had been fifty years earlier.[13]

This dramatic reduction in numbers clearly corresponds to the change in conditions that came with the transition from imperial to national rule. Here, a widespread, if paradoxical, dynamic was at play: minorities found that they enjoyed greater protection under imperial law, under which they could negotiate special sets of privileges, than they did under constitutional democracy, which asserted the theoretical equality of all citizens, but in reality provided no mechanism for the special legal protection of vastly outnumbered groups. In the case of Corfu, the conditions of both British (1815–64) and Greek (1864–) rule, for instance, were far poorer for Jews than during the Venetian and French periods. At the end of Venetian imperial rule, the local Christian community, long resentful of the "protected" status of the island's Jews, used the new political and legal circumstances to capitalize on its numerical superiority in a way it had been unable to do in the imperial context. In other places, too, where Jews were regarded with particular resentment as more "privileged" than Christians, the postimperial moment would bring a burst of intercommunal conflict, most notably in Salonika after 1912, when the Ottoman Empire lost the city to the Greeks in the Balkan Wars. But while in Salonika Jews had a near plurality in numbers, and to a degree could resist the growing Christian antagonism, in places like Corfu, where Jews made up only a small minority of the total population—in Corfu's case, around ninety-five thousand in 1907[14]—the response was rapid emigration, largely to towns on the mainland that were still under Ottoman control, such as Larissa and Jannina, but also to the Greek capital, which as Greece expanded was gradually becoming a major economic center. Meanwhile, in places where the Jewish population was poorer and not regarded by its Greek neighbors as particularly privileged—here, Jannina is the best example—the imperial/national transition was marked by fewer inter-communal outbursts.

With the switch from Venetian to French rule (1797–99; 1807–15), Corfiote Jews were granted citizenship along with the rest of the population; in the eyes of the law, all residents of the island enjoyed the same status regardless of faith. Jews and Christians alike were unhappy with the development—Jews because of the loss of privileges set by ancient precedent, and Christians because they felt that the Greek Orthodox were the rightful leaders of the island. In 1808, the French had to issue special legislation forbidding the harassment of Jews and in effect reinstate some of the protections of the Venetian period.[15] The downward trend continued during British rule (1815–64), when Corfiote Jews, though in theory equal to Greek Christians, were subject to more ill-treatment than they had been at any time in their past as old imperial protections faded away entirely. In practice, "equal rights" meant far less protection for Jews. As a reporter for the *Jewish Chronicle* put it in 1858, "The position of the Ionian Jews is that of the man between the two stools. . . . The Jews only exist to give sport to their Gentile task-masters."[16] During British rule, the incidence of anti-Jewish outbursts rose dramatically; there are widespread accounts from the period detailing the desecration of cemeteries, the disruption of Jewish funerals by jeering onlookers, and the escalation of the Easter-time physical assaults that occurred annually on the island until World War II.[17] The British observer George Orkney's *Four Years in the Ionian Isles* gives a vivid description:

> It is certainly curious that the Modern Corfiotes appear to cherish against the memory of Judas, a hatred greater than can be traced in the custom of other Christians. . . . [Just] for this occasion all broken or cracked earthenware are carefully preserved throughout the year. The supposition is that the good Christians are stoning in imagination the traitor Jew. . . . [On Easter eve, I] walked out at ten in the morning to see the follies. . . . At half past . . . , bang went the gun; and immediately from the tops of the windows of the houses down came the usual crash of crockery. . . . The Jews are not visible anywhere lately out of their own quarters. In most parts of the town they would now consider themselves unsafe.[18]

Rabbinic documents also testify to the lengths to which Corfu's Jews had to go to protect themselves, "since here in the island of Corfu most of the years [at Easter] our enemies confine us and prohibit us from leaving our homes because of the killing of their god."[19]

But as was the case with the Pacifico affair, anti-Semitism on Corfu as elsewhere was only one thread in a complex web of social and cultural factors, some of which had as much to do with internal communal dynamics as they did with intercommunal conflict. In the case of Corfu, two were critical: first, the fear on the part of Jewish leaders that proximity to Christians would lead to the corruption of the Jewish community, and second, the tremendous tensions between the two main factions that together made up Corfu's Jewish population.

A religious ordinance (*takkanah*) from the early 1700s, first issued by the Rabbi Menachem Vivanti, addressed the unfortunate overlap between the Greek Orthodox and Jewish liturgical calendars. The festival of Shavuot, which falls on the forty-ninth day after the beginning of Passover and commemorates Moses's receiving the tablets in Sinai, often coincided with the Orthodox Feast of the Ascension, which falls on the fortieth day after Easter. As was common throughout the Greek Orthodox world, the Feast of the Ascension was marked by a great outdoor festival (*panigýri*) with much eating, drinking, and merrymaking. The Jews of Corfu apparently marked Shavuot in a similar way, even celebrating in the same location where the Christians held their festival. Reissued by Rabbi Yehudah Bivas (perhaps the most famous Corfiote rabbi) in the late nineteenth century, the ordinance reminds the Jews of the impropriety of behaving in what might be construed as a Christian manner:

> During the days of Shavuot, some of the families with their sons and women and their children leave early in the morning and go outside of town, with a lot of food, and they park themselves in the same place where the Christians do their festival every year, and where they celebrate their feast, which is the Ascension. . . . And in that place there is eating and merrymaking and dancing and frivolousness and light-headedness. At that same field, at that same place, just as they [i.e., the Christians] do. And they are drinking there for at least the entire day. [But] with their zealousness for God he [Rabbi Bivas] and [now] also Rabbi Menachem Vivanti has been able to uproot this *'avodah zarah* [evil doing] from our land, praise be to God.[20]

The problems posed by living in close proximity to Christians came not simply from the Jews' seeming alien to them but also from the Jews' behavior seeming too much like theirs. Such edicts appear to have been widely ignored; this one was issued at least two times in Vivanti's tenure alone. An earlier version scathingly accused the Jews of "stuffing their stomachs" (*litsvot beten*) and drinking to the point of light-headedness (*kallut rosh*) beside the fountain of Kardaki—one of the most famous sites on Corfu, famed since antiquity for the magical properties of its waters. Set in a beautiful grassy spot within easy walking distance from the capital, with commanding panoramic views, the fountain had held religious significance for millennia, so it's not surprising that the Jews, like the Christians of the island, chose to celebrate there. Corfu's Jewish leaders feared, though, that such behavior was unbecoming to Jews and might be misconstrued as a form of participation in the Feast of the Ascension of the Virgin. The physical proximity of the Jewish and Christian communities, and the overlay of the Jewish and Christian liturgical calendars, created anxiety both within and between the two groups.[21] Similar anxieties are found in rabbinic literature from around the region. In the early 1500s, for example, the famous Cretan rabbi Eliyahu Kapsali complained that on the Sabbath his congregants behaved like Christians, not attending to services in the synagogue, but instead "go[ing] through the vineyards to the beach, and . . . head[ing] to the boats [to] seek only clowning around, and . . . things that please [their senses]."[22] Communal difference was something that both Jews and Christians had an investment as well as interest in preserving.

But if the anxieties of Christian-Jewish cohabitation were great, greater still in many places were the tensions within the Jewish community itself. Here again, Corfu gives an extreme example. Across the south Balkans, most of the region's Jewish communities were marked by the tension between Greek-speaking Jews, who claimed historical and cultural primacy, and Sephardic newcomers, who were generally wealthier and more cosmopolitan.

In the case of Corfu, a long history of multiple cultural influences, notably Italian and Greek, had over centuries led one group of Jews resolutely to define themselves as Greek, in distinction to other Jews on the island, who came from varied European backgrounds and were known as the Italian or Apulian (*Pugliese*) community, but which by the nineteenth century was largely Sephardic. This Greekness—expressed, paradoxically, by the Italian term (*Griego*)—was understood in an entirely Jewish context and had no connection to the Greek Orthodox population whatsoever.

Corfiote Jewish Greekness revolved around liturgy and religious literary tradition; some of the oldest existing prose composed in demotic Greek are Jewish songs and a version of the Book of Jonah, which the Jewish Greek community of Corfu traditionally read aloud on Yom Kippur.[23] At issue too were claims of "purity"; members of the Italian community complained bitterly of the Greeks Jews and their "pretension of wishing to appear as more ancient settlers than we Italians of Corfiote origin . . . [and] calling themselves the 'native Corfiote Hebrews' and wishing that we be styled 'foreigners.'"[24] While Corfu's circumstances are unique in Greek Jewish history, the tension between Greek-speaking indigenous Jews and "foreign" newcomers is typical of the region's communities.

The long-standing conflict between Corfu's Greek and Italian Jewish communities is the centerpiece theme of centuries-worth of petitions, letters, and rabbinic commentary arguing that certain privileges were the exclusive provenance of one group and were not to be shared by the other. That is, not only did the Christian community resent Corfu's Jews for their perceived privileges; each Jewish community begrudged them of the other. Indeed, Corfu's Jews were so divided and strife ridden that to refer to them as one community is to impose on them a unity that did not exist until well into the twentieth century. Unprecedented in Jewish history, the members of one group consistently and as a matter of policy sided with the Christian authorities against members of the other. Under Venetian rule, Corfiote rabbis fought unsuccessfully, for example, to prevent the members of one or the other group from reporting to the Venetian authorities on the financial status of members of the other congregation. In the effort to make members of the rival community pay higher taxes, members of one congregation (*kahal*) were snitching on members of the other. Rabbis cautioned, "In the name of Gabriel the angel of God . . . and in the name of God . . . [no one is] to advise [the authorities] by way of cunning and tricking, to help them take money from the Jews, and they are not to gossip with any of the authorities [*sarim*], and not to say, 'so-and-so' can pay the tax [*tanza*]." Violators were threatened with expulsion from the community (*herem*) and cautioned only to speak of such matters between Jews (*yehudi le-yehudi*).[25]

The astonishing tendency of the Corfiote community to turn to Gentile authorities to adjudicate their own disputes continued to the modern period. Both the Venetian and Corfiote Jewish records show the recurrent involvement of Venetian courts in the adjudication of Jewish conflicts.[26] In 1731, the sages of one of Venice's seminaries (yeshivas), the yeshiva Klalit, admonished the Corfiote Italian community for its disrespect for rabbinic authority and tendency to settle disputes not according to Jewish law (*din torah*) but according to Gentile law on the island—what the sages referred to as *dinim zarim* (Gentile law; lit., "pagan" or "evil" law).

> Our loins are filled with shuddering, we were seized with contractions, like those of a woman in labor, it terrorized our joy, because we heard the noise of a brouhaha, we heard the noise of war in your holy camp . . . a noise that profanes the name of the lord and honors the names of the idols. To follow other gods, to worship them with devotion, and to burn incense for their nostrils. And to settle financial disputes with strange laws. . . . While the Torah mourns over this— that they play in the yards and the castles with the children of Gentiles. . . . Please, our brethren, people of our redemption, pay attention to what you are doing! Will you always live by your swords? . . .

Drive the hate from your hearts, and with one heart, as in one man, plead your cases and your quarrels with truth, with law, and with justice so we shall never hear the noise of quarrels again. And send [your disputes] to any yeshiva you want, either in the kingdom of the Turks or in Italy, and the rabbis will tell you what to do according to the Torah.[27]

Despite such efforts for unity, however, the practice of siding with Gentile authorities continued. A century later, in 1831, the Italian synagogue hired a new rabbi, Yehuda Bivas.[28] Bivas was to lead daily prayers in the Italian synagogue, oversee the students in the Talmud Torah, visit the sick and dying ("except those with a contagious disease"), and give a sermon (*derash*) on important holidays. He was emphatically not, however, to introduce any liturgical or practical innovations: "The rabbi will follow all the customs [*minhagim*] of today, without any change."[29] The community didn't want any meddling from outsiders, especially after its long history of conflicts with the Greek Jews.

Well into the nineteenth century, Corfu's community was so divided that Christian authorities were regularly asked to adjudicate intercommunal disputes. In autumn 1837, Pellegrin Richetti, an aristocratic Venetian Jew who had come to the island, expressed his desire to serve as a cantor. Rabbi Bivas rebuffed him as a foreigner. Richetti went promptly to Don Vincenzo Donadi, governor of the island, to complain that Bivas was running "an unauthorized house of prayer in his [own] home" and intentionally setting members of the community against one another, "spoiling the brotherhood that is meant to exist between Jews." The governor contacted Bivas; in turn, the rabbi sent back a list of counterarguments, and threatened to sue Richetti for slander and condemned his character.[30]

The divide between the two communities mirrored Corfu's geographic location between Venice and the Ottoman Empire.[31] Nineteenth-century visitors described the island as the crossroads between the West and the East: "Let the traveler [to Corfu] luxuriate in his last taste of European life, remembering well the difference between the Ionian Islands and the rest of Greece. . . . Strange tongues greet his ear and Eastern costumes delight his eye, side by side with English and Italian accents and the familiar garb of the West."[32] Similarly, Corfiote Jewish practice, too, reflected a marked hybridity; until the beginning of the nineteenth century, the Italian synagogues of Corfu used a mix of languages, and prayer books included words in Greek, Ladino, Hebrew, and Italian (Apulian dialect). At the same time, Corfu's divisions remind us to be cautious of tidy communal categorizations. Jews, no less than any of the region's inhabitants, were shaped by various historical and cultural influences. In the case of Corfu, the division of the Jewish population into two communities roughly reflects the two main cultural influences present on the island as a whole and can be loosely seen as replicating the Catholic/Orthodox divide within its Christian community.

Corfu's Greek Jewish community was the older, and far more xenophobic; throughout its long history, it consistently refused to admit any newcomers and repeatedly asserted its specific privileges vis-à-vis the Venetian government, to which it appealed to prevent other Jews on the island from sharing the same rights. Here, a fairly widespread dynamic was at play. The resistance of the Kehila Griega to sharing its privileges with the Kehila Italyani (or Pugliese) reflects the Romaniote desire, common throughout the region, to retain its unique character. Corfu's Greek Jewish community managed to maintain its juridically superior status for a long time. But its insistence on its historical, cultural, and legal integrity gradually eroded its stature and power. The Greek community's hesitance to accept newcomers meant that its rival community, the Italian, or Apulian, grew steadily in numbers while the Greek didn't.[33] While the Italian community, which like the Greek, had existed since prior to Venetian rule, grew to encompass not just Italian Jews from the mainland but also Spanish, Portuguese, and Ashkenazic refugees, the Greek did not change in size or makeup.[34]

ROMANIOTES AND SEPHARDIM

As Corfu vividly exemplifies, in looking at Greek Jewish history what one is really exploring is multiple histories, of multiple peoples, and tracing the story of how they shakily emerged as a unified, homogenized entity in the twentieth century. In this progression, a key issue throughout the region was the conflict between indigenous, Greek-speaking Jews and largely Sephardic newcomers, and the ways in which that was negotiated against a complex broader political backdrop. In almost all places, the conflict was ultimately settled in favor of Sephardim, but nowhere was it settled definitively or easily. Tensions within and between different Jewish communities were another variable in the degree to which the impact of the transition to Greek rule was felt.

The dynamic of the Romaniote-Sephardic conflict, with gradual Romaniote diminution and Sephardic expansion, was most famously present in Salonika, where Sephardim early on subsumed the Romaniote community. As early as 1509, a Sephardic ḥakham (rabbi; lit., "wise one") had rejoiced,

It is well known that Sephardic Jews . . . comprise the majority here, may the Lord be praised. The land was given uniquely to them, and they are its majesty, its radiance and its splendor, a light unto the land and unto all who dwell in it. . . . For all these places are ours, too, and it would be worthy of all the minority peoples who first resided in this kingdom to follow their example and do as they do in all that pertains to the Torah and its customs.[35]

But even the story of conflict between Sephardim and Romaniotes is more complicated than it initially appears. The central event in early modern Jewish history—the expulsion of the Jews of the Iberian Peninsula—wasn't experienced as the single, dramatic rupture that modern Jewish memory so often describes it to be. Instead it was a series of events, a string of migrations, first from Spain, then from Portugal, Italy, and at last Navarre, to North America, northern Europe, North Africa, Anatolia, and the northwesternmost provinces of the rapidly expanding Ottoman Empire. These various groups, now collectively designated as Sephardim, first had to negotiate their own differences. In the 1560s, Rashdam (Samuel, b. Moses di Medina, 1505–89) complained of the proliferation of sects that the varied refugees had brought with them to the Ottoman Empire:

In this city of Salonika, where the law states that each man must remain true to his kahal, sometimes a few leave the kahal to worship in another synagogue, and no one stops them. Even today, in two prominent and separate congregations from Spain, in the holy congregation "Gerush Sefarad," may the Almighty protect them, a few members left because of disagreements and established their own synagogue, and a short time ago, some members of the holy community of Shalom left to worship in the synagogue of Provincia, a place neither they nor their fathers ever knew or imagined.[36]

As these various traditions were brought together under the hegemony of a few leading Sephardic groups, Romaniotes, too, gradually changed their liturgy and customs, adopting Sephardic ones in their place.[37] This slow and incremental transformation left only a few "pure" Romaniote communities, in places like Jannina, Crete, and Halkis. In some areas, like Corfu, Volos, and Larissa, Romaniotes and Sephardim coexisted.

In places where Romaniotes managed to remain dominant and distinct, the expulsions from Iberia nevertheless had a worrisome effect. In sixteenth-century Crete, for instance, the Jewish community felt beset by problems, many of them imported along with Jews arriving after the expulsions, and much like Corfu's Kehila Yavanit, resisted the encroachment of Sephardic newcomers. Even in Romaniote strongholds like Jannina, Romaniotes struggled to maintain the cultural upper hand, as the rabbinic principle that Jews follow the customs of the lands to which they move was ignored by the Sephardim, who brought their own customs with them and imposed them on the Romaniotes.[38] Ultimately, even in Jannina Romaniotes succumbed to Sephardic pressure to the extent that they started sharing Sephardic liturgical usage and depending on Sephardic rabbinic rulings, although the two communities remained officially separate. By the seventeenth century, the Salonikan rabbinic establishment had such hegemony in the region that *responsa* literature dealing with Jannina was written by Salonikan Sephardic rabbis, not by Romaniotes. Other ancient Romaniote communities—Veroia, Florina, Halkis, Kavála, and Kastoria in the north, as on Crete to the south, the Ionian Islands to the West, and Rhodes to the southeast—were subject to the same pressures, and because of their smaller size, were more fully and rapidly assimilated to Sephardic practice and tradition during the seventeenth century.[39]

During the early modern period Sephardim established networks for trade, scholarship, and education that stretched from European ports in the west to Palestine in the east, and from the Balkans in the north to the Maghreb in the south, and Romaniote Jews could not afford to be excluded from them. Over time, Greek-speaking Romaniote communities that resisted assimilation to Sephardic tradition found themselves increasingly isolated, with neither a numerical majority (*rov minyan*) nor a greater wealth (*rov binyan*). What Romaniote communities gained, then, in terms of an identity resting on notions of cultural elitism, integrity, and alleged purity was amply outweighed by their shrinking influence in terms of size, overall wealth, and cultural contacts around the region.

The establishment of Greece and its gradual expansion over the next century brought about, to an extent, a reversal of a trend that had dominated for centuries. With the integration of large swaths of northern territory into Greece, Romaniote Jews, for the first time in centuries, had a certain advantage vis-à-vis their Sephardic coreligionists. Their knowledge of Greek, contacts in Old Greece, and greater level of integration into Greek culture all made it easier for them to negotiate the new social and legal circumstances presented by the territorial expansion of the early twentieth century. Sephardic Jews increasingly saw the ability to speak Greek as a critical advantage, and the pro-Ladino chauvinism that had dominated for centuries gradually began to give way. While before integration into Greece it was the Ladino speakers who were able to capitalize on far-flung transnational contacts around the region, and could move between the Jewish worlds of Greece and the Ottoman Empire with ease, after integration it was the Greek-speaking Jews who were more readily able to move between two worlds—in this case the Jewish and Christian ones.

One example among many is that of Giomtov Giakoel, a Sephardic native Greek speaker who was born in Trikala in 1898. After taking a law degree in Athens, Giakoel moved in 1923 to Salonika, where he opened a law office that serviced the city's Jewish population. His many contacts with Old Greece, his Athenian education, and the fact that Greek was his native tongue all contributed to his having an extremely successful practice in Salonika, where many Jews needed legal help as they tried to negotiate the new legislation and social ambiance that confronted them after unification with Greece in 1912–13, and again during urban reconstruction in the early 1920s.[40] In Thessaly, which became part of Greece in 1881, for three decades Jews were geographically sandwiched between a Sephardic, Ottoman imperial orbit to the northeast and a Greek Orthodox sphere to the south. Jews, like Giakoel, who were bilingual (in Ladino and Greek) found that their facility with both languages gave them an advantage; among the younger generation, it was not uncommon for Jewish businesses to establish office branches in Salonika and Athens, and work in bilingual trades that made it possible for them to serve both Ladino and Greek speakers.[41] At the turn of the nineteenth century, as Greece continued its ineluctable expansion, outlying bilingual Jewish communities (like Trikala) suddenly became indispensable to Salonika, which had hitherto served as the pole star around which all other Sephardic communities in the region orbited.

During the early modern period, the dominant influence on the region's Jewish communities had been Salonika's rise to primacy and the concurrent hegemony of Sephardic practice. As the territories that made up New Greece undertook the transition from empire to nation in the nineteenth century, the driving force for change was the increasing consolidation of Greek nationalism and the rapid growth of Athens from a small town to a major urban and economic hub. The very factors that earlier had worked to the advantage of Sephardim—their cultural and linguistic uniqueness, far-reaching scholarly and economic connections, and large numbers—would, in the context of the Greek state, set them apart and put them at a disadvantage. Conversely, Romaniotes, who for centuries had lived in greater isolation from the broader Jewish world, would find the process of becoming Greek a somewhat less difficult one.

In Kolettis's vision of the Great Idea, Greeks were Greek no matter where they lived—whether within the borders of the core Greek state (Old Greece) or the territories still under Ottoman control that would soon become part of it (New Greece). Much has been made of the expansionist and theological dimensions of the Great Idea. Most fundamentally, though, it can be read as the quintessentially transitional concept: a configuration of Greece as a space that was at once national and imperial. In the uneven transition from imperial subjects to sovereign nation, many Greeks understood their identity with both the national language of citizens and individuals, and the imperial one of religious collectives. During the long century of Greek territorial expansion, Jews, too, conceived of themselves at once in imperial and national terms.

An observer of the battle for Jannina, in 1913, famously observed, "The city that went to sleep Turkish and Ottoman woke up Greek and Christian!"[42] In lived terms, however, the transition from imperial to national territory was a gradual and awkward one, and it heightened long-standing tensions both between and within different religious groups. In the context of the aggressive Greek expansionist policies of the nineteenth century, the constant tension between the size of the actual Greek state and the Greater Greece of the imagination drew clear battle lines between those sympathetic to the Greek cause and those who opposed it. Jews, in particular, were regarded as suspect in their loyalties, since in territories not yet a part of Greece, Jews for the most part—but not uniformly—hoped that Greek expansion would be stopped in its tracks. The attitude of Jews outside of Greece—most notably, those in Salonika—had spill-over effects, compromising the security of Jewish communities within Greece. During expansionist offensives, Greek Jews made efforts to establish their loyalty to Greece and disassociate themselves from the pro-Ottoman demonstrations of Jews outside of Greece. When the Greco-Turkish War broke out in 1897, for example, a number of Jews (largely Romaniotes) in Old Greece (from Patras, Corfu, Zákinthos, Volos, Athens, and Halkis) took part, fighting on the Greek side, partly to demonstrate their loyalty as Greek citizens. Meanwhile, on the Turkish side some of the most ardent defenders of the Ottoman status quo were Sephardim. The Greco-Turkish War is perhaps the first instance in the modern period of Jews fighting other Jews in the context of national wars. This circumstance derived precisely from the fact that the imperial/national transition did *not* happen overnight, and reflects both the uneven transition from nation to empire and the concomitant circumstance that political attitudes did not inevitably fall out cleanly along communal lines.

As the core territories of Old Greece expanded to the north and east, the members of numerous small Jewish communities, scattered in pockets across the region, became citizens of the Greek Kingdom. While the transition was hugely significant to these individuals themselves, the absorption of these small communities had only a modest impact on Greece. With few exceptions, notably Jannina and Corfu, no one Jewish community numbered more than a thousand. When Volos became part of Greece (1881), about three hundred Jews lived there. Trikala (1898) was home to five hundred and Veroia (1912) perhaps a hundred more than that. On Crete (1913), there were about four hundred Jewish residents across the entire island. After eight decades of expansion, on the eve of the Balkan Wars (1912–13), Greece's total Jewish population was well under ten thousand, out of a total population of close to three million. And while Greece's Jewish population slowly climbed in numbers, the Jewish population of the region experienced an overall decline; that is, as territories became Greek, the tendency was for their Jewish population to contract with emigration. Within Greece, the only city that saw a steady strengthening of Jewish life was Athens, where the Jewish population grew from a literal handful to about twelve hundred in the first century of Greece's existence.

The decades preceding the Greco-Turkish War had seen a gradual consolidation of Greek Jewish life, particularly in Athens. In the context of nineteenth-century fledgling Greek Jewish identity, Romaniote Jews emerged as dominant. In the 1880s, the Greek state had for the first time recognized the legal status of Greece's Jewish communities—at that time almost all Romaniote. This had the effect of further drawing Romaniotes into the national orbit and away from an Ottoman imperial one. It also aided the consolidation of Athens as a Jewish, Romaniote center that was increasingly a counterbalance to Sephardic, Ottoman Salonika. As anti-Semitic violence popped up around the country in the wake of territorial unification, the Athens kehila emerged as a place of shelter. While in the wake of independence Jewish refugees had coalesced around Sephardic centers within the Ottoman Empire, increasingly during Greek expansion they moved to Athens. The Athens kehila extended assistance to Jews on Zákinthos and Corfu after they were unified with Greece, and absorbed many Sephardim from Thessaly during and after the Greco-Turkish War. From its inception, the Athens community had been made up of refugees and immigrants from an array of backgrounds. This helped Greek culture assert itself as dominant: while in Salonika Sephardic culture had a monopoly on Jewish life, the piecemeal nature of Athens' Jewish community greatly facilitated its integration into the dominant national culture. While Greece was defeated in 1897, the Greco-Turkish War nevertheless prompted many Jews to leave Greece's northern borders and head to Athens. Greece's constant military pushes into the north made it clear to even some Salonikans that their future lay with Greece. For the first time, rather than leaving Greek territory for the region's de facto Jewish capital, a slow trickle of Jewish migration began in the opposite direction as some Salonikans moved to Athens.[43] The small Athenian community, at the turn of the century made up of no more than four hundred Jews, saw a modest swell in numbers as Salonikans arrived.[44] With the expansion of Greek borders in the Balkan Wars fifteen years later, others followed them. Culturally, too, there was a reversal of past trends: while for centuries Jewish life in the south Balkans had assimilated to Sephardic Jewish tradition with the arrival of Sephardim, in late nineteenth-century Athens arriving Sephardim gradually adopted Romaniote culture and put aside Ladino for Greek.

Until 1912, Greece was home to no more than ten thousand Jews. With the conquest of Salonika in the Balkan Wars (1912–13), Greece's Jewish population would grow by seventy thousand. Yet culturally and historically speaking, the Jews of Salonika were the least Greek of any of Greece's populations. With a strong identity as Ottoman subjects, their own language, and the sheer force of numbers, Salonika's Jews were at once the least Hellenized and the most dominant of the various groups that together made up Greek Jewry.

As a broad generalization, Sephardim generally and Salonikans in particular suffered from the expansion of Greece, while Athens and Romaniotes benefited. This dynamic accelerated when, in 1904, Athens' total population outsized that of Salonika for the first time. The members of Salonika's Jewish community would pass from being a powerful and influential part of a multicultural multireligious empire to being unwanted minorities under the rule of a Hellenizing nation-state. In contrast, Romaniotes found that their fluency in the Greek culture and language gave them a huge advantage in the increasingly Greek New Greece. Now, it was they who were more readily able to negotiate privileges; in 1913, for example, the Athenian Jewish newspaper *Israilitikí Epitheórisis* (Jewish Review) announced that "the increase of the Greek Jewish population . . . [had inspired] the Kehila of Athens" to ask for special legislation allowing for Greek Jewish education.[45] Within four years, the law 1242 was enacted, requiring the appointment of Hebrew teachers in city schools with classes including more than ten Jewish students. As Athens Jews successfully negotiated with the Greek government, Salonikans at the same moment fought to push back a host of new regulations designed to give economic advantages to Greek Orthodox residents.

With the 1912 annexation of Salonika to Greece, followed two years later by its formal incorporation, the comparative Greekness of Athens would be underscored. Salonika was indisputably the more important Jewish center, but it was also, as Nikos Stavroulakis puts it, "a closed world, even to other Jews." Athens was increasingly the city of choice for a new, more Hellenized and secularized generation of Jews. While Salonikan Sephardim were undergoing a painful, forced, and only partially successful process of Hellenization, Romaniotes and younger Jews from around Greece looked to Athens as a place where they could both "assert their Greekness" and find a substantial number of coreligionists.[46]

But it was only with the incorporation of Salonika into its national space that Greece would become a truly Jewish country. Only then would the term Greek Jew refer to large numbers of people. Here, then, is one of the greatest paradoxes of the history of Greek Jews: it is a history, to a large extent, of a group of people who badly wanted to be something other than Greek, and who came only in the last decades of their existence to think of themselves as such.

PART TWO

The "Sephardic Republic": Salonika to 1923

Chapter 4

SALONIKA TO 1912

> Let me tell you something about Salonika, who these
> Jews are. Because in understanding Salonika, you will
> understand, I think[,] the picture of Greek Jewry.
> —Israel G. Jacobson, address to the National Council
> of Jewish Women, February 27, 1946.

GREEK JEWRY, in the modern sense of the term—that is, Jewish citizens of
the Greek nation-state—refers overwhelmingly to the Sephardic descendants of the Iberian expulsions, who settled in the Ottoman city of Salonika (Selânik) during the sixteenth century. On the eve of the Second
World War, eight out of nine Greek Jews were Salonikan Sephardim. Yet
Salonika's Jews were Greek only for a brief time: from the 1912 annexation of the city in the Balkan Wars until early spring 1943, when the
German occupying forces began their process of mass Jewish deportation
to the death camps of northern Europe. And while it is Salonikan Sephardim who together make up the vast bulk of Greek Jewry, until 1912 there
was very little indeed about them that could be considered Greek, in any
sense of the term. In significant ways, then, one whole swath of the Jewish
history of Greece is a fleeting, elusive, and phantom one. It is as much a
counterhistory, a history of what might—should—have been, as it is of
what was. As lived history, it is a largely diasporic one: the history of
Greek Jewry and its descendants beyond the borders of Greece. Within
Greece, it is the history of the small group of survivors and the community
they have built in the wake of unthinkable destruction.

THE *"JERUSALEM OF THE BALKANS"*: YERUSHALAYIM DE BALKAN

To the extent that to understand Salonika is to understand Greek Jewry,
it is crucial first and foremost to understand that although Greek Jewish
history is to a large extent the history of Salonika, the history of Salonika's
Jews in actual fact has very little to do with Greece. Salonika's history, so
eloquently told in Mark Mazower's recent *Salonica, City of Ghosts*, is an
Ottoman history, and the history of its Jews is an exilic one.[1] Salonika's
Jews were, above all, a people with a profoundly exilic identity, an identity

powerfully shaped by the memory of Sepharad and the expulsion from it. Salonikans developed a deep attachment to the exilic condition and the place of exile—the city of Salonika. Indeed, with the near destruction of Salonikan Jewry in World War II, Salonika itself would replace Iberia as the longed-for topos of origin. For centuries, from the fifteenth century on, Salonika's Jews named their synagogues after locations in Spain and Portugal, sang songs of Iberia, and "remembered" home in countless religious and cultural rituals. As one traditional song, recently rerecorded, asked: "Onde esta la yave ke estava in kashun? / Mis nonus la trusherun kon grande dolor / De su kaza de Espania / Suenios de Espania" (Where is the key that was in the drawer? / My forefathers brought it with great pain / From their house in Spain / Dreams of Spain).[2] In Israel today, the descendants of Salonika's Jews now attend synagogues named for Salonikan neighborhoods and sing songs that memorialize their Ottoman hometown. Yehuda Poliker, for example, a contemporary Israeli pop singer of Sephardic descent, sings plaintively of "*Glykiá mou Saloníki*"— my sweet Salonika—and laments his *xenitiá*—his condition of exile from his Greek home.

The "sweet Salonika" Poliker and other Sephardim remember was an Ottoman city, built on the twin foundations of Western Catholic discrimination and Ottoman Muslim expansiveness and territorial expansion. Its inhabitants had their origins in a series of uprooted and displaced groups: Ottoman slaves, refugees, merchants, and travelers. The Spanish and Portuguese expulsions of the late fifteenth and early sixteenth centuries brought waves of Jewish migrations from Iberia, France, and what today is Italy to the Ottoman Empire. Salonika's Jews forged a unified Sephardic identity out of an extremely diverse group of Jewish refugees, and created a unilinear, locative history out of what was actually a remarkably variegated and dislocated one.

Salonika's rabbinic establishment provides a good example: the rabbinic authority for which Salonika became famous was forged in the confusing social and religious circumstances created by the expulsions, which threw together Jews—largely Sephardim, but Ashkenazim and Romaniotes too—of widely ranging backgrounds. While by the nineteenth century the Salonikan rabbinic establishment was known for its conservatism, in the sixteenth and seventeenth centuries it was innovative, even radical—a response to the wide range of peculiar halachic problems created by the expulsions. Most pressingly, many of the émigrés—who continued to arrive until well into the seventeenth century—were Portuguese *conversos*, Jewish "converts" to Christianity, converted by Portuguese monarchical decree in 1497, whose families had lived for two or three generations as Christians, and now were leaving Portugal to escape the Inquisition and find a place where they could legally return to Judaism.

These converted Jews posed a tremendously difficult challenge to the rabbis who received them in the western Ottoman lands and had to determine how best to accept them back into Judaism, and foster and monitor their reincorporation into the Jewish kehila, in the broadest sense of that term. More than any other factor, the presence of these converts led to the development of an unusually creative and vibrant rabbinic tradition.

The vexing legal and cultural problems attendant on the troubled status of conversos gave rise to a huge, unique, and rich body of legal writing on the adjudication of *teshuvah* (return to Judaism; repentance).[3] Virtually all rabbis of the period had to confront the specific problems that came along with the mass immigration of "returning" Jews who had lived all their lives as Christians.[4] The urgency of providing mechanisms for their reentry into Jewish life made for a flexible rabbinic establishment. For the most part, Salonika's *kehalim* (congregations) were determined to make the process of teshuvah as smooth as possible for the arriving conversos.[5] Indeed, in many cases the reversion to Judaism was automatic and required no formal adjudication whatsoever. As one responsum, written in the 1540s by Rashdam, a leading Salonikan jurist, put it, "[Regarding] all conversos who are coming to do teshuvah: [we must] consider that he whose father is of Israel [i.e., Jewish], his mother also must be assumed to be not a Gentile."[6] That is, if a converso who undertook teshuvah identified their father as Jewish, that was enough to make one Jewish. In the same vein, another Rashdam responsum stated explicitly that even the children of conversos, born and raised as Christians, were to be considered kosher Jews: "Those who are children of conversos are called 'Israel,' even after several generations."[7] Rabbi Yosef Ben-Lev (Rival, 1500–1586), another Salonikan halachic figure, was even more radical in his approach to the question of teshuvah. Rival argued that people who had not come immediately to the Ottoman Empire but rather had lived for some time as Christians in other lands were not to be judged harshly for their initial decision to continue living as Christians; in their hearts, he contended, they had always been Jews. "Those [converso] Jews who left their residence in the kingdom of Portugal on their way to Turkey— even if some of them end up residing in Ancona and Flanders [where they continued to live as Christians]—. . . one can say that they are kosher Jews, because in their hearts they have considered teshuvah."[8] On the topic of Jews who continued to use their non-Jewish Portuguese names, Rashdam was similarly flexible. While it would be an act of piety if these individuals were to use their Jewish names instead, they weren't obliged to, and couldn't be regarded as "less Jewish" if they didn't.[9]

Within two hundred years, by the eighteenth century, Salonika's rabbinic establishment was known as hegemonic, even in its rulings regarding Salonikans, much less Jews from outside. This change in orientations is powerful testament to the swift consolidation of Salonikan Jewish social and economic power over the course of the sixteenth and seventeenth centuries. But if legal consolidation ultimately won the day, Salonika's learned establishment never lost its early traits of a diversity of opinion, vigorous internal debate, and an extraordinarily vibrant intellectual life. All were characteristic until the community's destruction in the Holocaust. While over time Sephardic Salonika would come to be regarded as a closed society, even to other Sephardim, internally it remained consistently diverse, argumentative, and intellectually alive.

The cultural homogeneity and rabbinic diversity of the sixteenth century, which gave way over the course of the seventeenth to increasing consolidation, was the crucible for the birth of a rabbinic and scholarly tradition that was arguably unsurpassed in early modern European history. It worked in tandem with the development of an Ottoman Jewish economic sphere of which Salonika was the center to establish Salonika's primacy as a Jewish city. Indeed, for the last three centuries of the Ottoman Empire, Salonika reigned as what the Rabbi Moshe Almosnino (Moses Ben-Baruch Almosnino, 1515–80), the great Salonikan rabbi, political counselor, and sometime historian of the sixteenth century, called "*la republica sefardita*," the Salonikan Republic—an epithet that captures the sense of both the strength and autonomy that Salonika's Jews enjoyed for centuries.[10]

THE GRECO-TURKISH WAR

This story, though, is *not* "the story of Greek Jewry." It is the story of the most significant community within Ottoman Jewry, and is an Ottoman and Jewish but not a Greek one. The Greek story of Salonika began, tentatively, only in the nineteenth century, as Salonika's Jews warily looked on at Greek territorial expansion, well aware that Greece had set its sights on their city.

Immediately after the War of Independence, the Greek Foreign Office had established departmental divisions corresponding to the various Ottoman provinces that Greece aimed to conquer: Macedonia, Thessaly, Epirus, and Crete. From the start, the borders of the new Greek state were regarded as provisional, at least so far as the Greeks themselves were concerned. It was no secret that Greece aimed ultimately to include Salonika within its borders. By the 1890s, the question of what was to be done with Ottoman territory in the event of an Ottoman collapse was largely one of what would become of the city of Salonika. Various Balkan peoples had made dramatic moves to secure their own nation-states. Serbia (1815), Greece (1829), Romania (1878), Montenegro (1878), and

Bulgaria (1878) each in turn claimed independence. Each, like Greece, sought the liberation of its "unredeemed" members, using the goal as the basis for irredentist/expansionist agitation. In the case of Greece, virtually all other dimensions of nation building—the development of the economy, infrastructure, and government—were subordinated to the overarching goal of territorial expansion.[11]

All parties were particularly interested in Macedonia, an imprecisely delineated chunk of land overlaying parts of European Turkey, Greece, Bulgaria, and Albania, of which Salonika was the capital and the only city of major economic significance. While other territories—Crete, Thessaly, and the Ionian Islands—that had joined Greece had populations that were overwhelmingly Greek Orthodox, in Macedonia there was a far more mixed demographic. Salonika itself was less than half Greek Orthodox. Thus, in the case of Salonika it was not the city's population that needed redeeming so much as the city itself. The Greekness of other territories has been asserted by reference to the religion of the inhabitants. In Salonika, other factors were at play. To Greek minds, it was unthinkable that Salonika would end up anything other than Greek: Salonika was larger and wealthier than Athens; its geographic location was key to both overland and sea trade; and all goods and commerce in the region had had it as a hub for centuries. In contrast to the classicism and heavily Europeanized Greekness of Athens, it was the site of the rebirth of "*Romiosýni*"—demotic, popular Greekness.[12] It was also regarded as "organically" part of the landmass that was Greece. Greece's main competitors were the Bulgarians and Serbs, but the desirability of the city was such that even as distant a suitor as Italy, at the time in a confident and modernizing mood, dreamed of including the city in its Illyrian rebirth.[13] The multiple and mutually exclusive claims of this host of suitors were a virtual guarantee of war. Greek claims to the city and the ideology of the Great Idea also meant that should Greece gain control of Salonika, its rapid Hellenization would be essential.

The southernmost territory claimed by Greece—Crete—also emerged as a critical factor in the events looming on the northern frontier. Cretan adventurers and volunteers had been prominent in irregular bands that fought against the Bulgarian *comitadjis* in Macedonia during the second half of the nineteenth century. They were violently opposed to anyone suspected of disloyalty to the Greek cause. Cretan Greeks embodied the revolutionary ideals of independence, valor, and a hatred of the Turk. Cretan uprisings in almost every decade of the late nineteenth century called for unification—énosis—with Greece. These revolts were awkward for Athens, which wanted territorial expansion to proceed on its own terms. The Greek citizenry was unhappy with the government's position; increasingly the cause of Crete and that of Macedonia became linked, and the monarchy was accused of treachery for not pushing rapidly forward with territorial expansion. What John Koliopoulos has aptly termed *"pallikarism"*—a Greek expansionist machismo fused with an aggressive nationalism—brought Greece to the brink of war by the late 1880s.[14] In 1896, the first modern Olympic Games were held in Athens. Greek national sentiment was at its peak. Greek insurgents were being brutally slaughtered by the Turks on Crete, and Bulgarian claims to Macedonia were increasingly assertive. The Ethnikí Etairía (National Society) clamored for the Greek government to preserve Greek honor. The press was filled with inflammatory articles demanding military action. Finally the government, already heavily burdened with debt and scarcely able to mount a military campaign, capitulated to public demand and declared war on the Ottoman Empire.

The Greco-Turkish War, which came to be known as "the Shame of '97," proved a complete disaster. Within weeks the Turkish army was marching on Athens, and the great powers once again had to intervene on Greece's behalf. Greece was forced to surrender. A huge indemnity was but one of the humiliations. Small portions of land were ceded to Turkey, an international finance control commission was set up to oversee the payment of Greece's massive international debt, and the Western degree of involvement in Greece's affairs, long a source of resentment, increased.[15] In the only positive feature of the postwar settlement, Crete was granted autonomy, and the Ottoman influence on the island was reduced to a nominal level. Over the long run Greece emerged quite unscathed from the loss, but at the time the psychological sense of dejection was vast. Almost immediately, the search for a scapegoat began. The easiest targets were the usual ones: the palace and outsiders. As Koliopoulos puts it, "It was pallikarism which led to the disastrous war of 1897 and [that] made most politicians and army officers blame only the palace and the foreigners for the disaster."[16]

It was within this broader and fraught context that Salonika's Jewish residents emerged as a plotline within the modern Greek historical narrative, albeit as an unwilling and unwanted one. Increasingly, the Greek press and public began to regard Salonikan Jews and Greek Jews—that is, those Jews already living within Greek borders—as functionally one and the same. The Salonikan Jewish response to the Cretan uprisings, for example, was followed closely by the Greek press both within and beyond Greece's borders. Thousands of members of the Salonikan Jewish community took to the streets chanting, "Todos muramos, las Creta a los grecos non la damos" (We would rather die than give Crete to the Greeks). At issue, obviously, was not so much Crete itself—hundreds of kilometers to the south and across the Aegean Sea. The real concern was Greek expansion and Ottoman collapse. Greek observers began to complain of the potential disloyalty of Greek Jewish subjects.

The Greco-Turkish War marked a tipping point in the fortunes of the ten or so thousand Jews within Greece, who until the end of the century seem gradually to have been settling into a normalized existence as Greek royal subjects. With the outbreak of the 1897 war, an episode involving Greeks, Turks, and Jews highlighted the awkward triangular relations of the three groups and demonstrated the degree to which Jews within Greece were increasingly under pressure as a result of the pro-Ottoman loyalties of their Salonikan coreligionists. In April, a group of Greek prisoners of war being led to the Salonika train station by Turkish guards was set on by a band of Jewish youths, who mocked them and, by some accounts, attacked them with a shower of stones. In today's parlance, the episode was clearly framed by "ethnic bias"—the Greek prisoners were mocked *as* Greeks. Less clear is the extent to which the actions of this one group of "louts," as one historian has termed them, were representative of the Jewish community as a whole.[17] Significant, though, is that from the Greek perspective, they were. An article in the Athenian newspaper *Akrópolis* described Salonika as a town in which "the Christians are persecuted more by the Jews than by the Turks," and depicted a "mania of persecution": the desecration of churches, the destruction of Christian property, and a state of fear among the Christians, who "do not dare to go outside lest they be pelted with stones."[18] The Greek press published a series of editorials on the "Pro-Turkish Stance of Some Jews in the Greek-Turkish War of 1897," describing the railway station incident as one manifestation of "community-wide" Jewish celebrations of the Greek defeat.[19] Other articles recalled that Jews had been sympathetic to the Turks in the War of Independence in the 1820s, during the Cretan uprisings, and again during the recent Greco-Turkish War. A picture began to emerge in the press of the perennially Turkophilic Jew—anti-Greek, disloyal, and untrustworthy.

Judging from the Jewish reaction, this seems to have been a new development. Throughout Greece, Jews felt compromised by the events and particularly their reception in the Greek media, which as the Corfu community phrased it in a letter of rebuke to the Salonikans, "seriously compromises our tranquillity and that of all our coreligionists throughout the kingdom." Increasingly, the pro-Ottoman actions of Salonikan Jews were held against Jews living in Greece. As the Corfiotes urgently put it to Salonika's leadership:

> You must take the most energetic measures to make our co-religionists there understand that every demonstration, every hostile act can do grave harm to their brothers in Greece by exciting the indignation

and hatred of the population against us; we are already afflicted
enough by our nation's disastrous defeat and by the sad consequences of this unfortunate war.[20]

For its part, the Salonikan community was so distressed by the possible
repercussions of the episode that it sent messages of warning to the leadership of Jewish communities around Greece and strenuously argued that
the acts of a few "low-class ignorants" in no way represented the sentiments of the majority of the Jewish population.[21] The British authorities,
to whom the Greek government had appealed in the matter, ultimately
drew much the same conclusion. The Greek government, in their view,
had grossly overexaggerated the episode.[22] But the damage was done.
Greece's Jews had established a tentative modus vivendi over the preceding years. With the Greco-Turkish War it was threatened, if not derailed—
a circumstance that reveals the extent to which imperial communal categories continued atavistically to cut across the new, nationalist boundaries
of the nineteenth century. The destabilizing effect of the Greco-Turkish
War and the "Salonika Question" on Greece's Jews is further reflected in
the early twentieth-century popularity of the Zionist movement within
Greece as opposed to the Ottoman Empire. A Zionist newspaper was
issued on Corfu in 1899, and Zionist organizations were founded in Larissa in 1902, on Corfu in 1906, and in Volos in 1910. Meanwhile, in
Salonika, the Zionist movement gained little ground.[23]

At first, then, the question of Salonika created a sense of pessimism
and insecurity that was felt far more acutely by Greek Jews affected *by*
Salonika that it was by Jews actually *within* Salonika. After all, when
the Turks defeated the Greeks in 1897, Salonika was still both larger
than Athens and the largest Jewish city in the world, in terms of the
percentage of the total population.[24] Sephardic Jews made up the plurality of its population, and would continue to do so until the population
exchanges between Greece and Turkey in 1923. The Sephardim remained in control of the city's most important businesses and its ports,
where Jewish stevedores were prominent, and that still closed for business on Jewish holidays and the Sabbath. Meanwhile, Jews in Greece
were a tiny minority, and for the most part were far poorer and with far
fewer social as well as communal resources than Jews in Salonika. Anti-
Semitic sentiment had far more dire implications for Jews within Greece
than for those in Salonika.

Within a decade, though, a strong sense of insecurity began to characterize the Salonikan community as well, as it became all but inevitable
that Salonika would end up in either Bulgarian or Greek hands. The idea
of Salonika as anything other than Ottoman seemed unthinkable to most.
"Salonika will become like a heart that has ceased to beat," wrote one

Jewish leader of the possibility. "It will become like a head that has been torn from its body."[25] Nineteenth-century concerns over the preservation of Sephardic cultural heritage gradually gave way to strategic attempts to protect Jewish dominance in Salonika's economy, and religious debates were increasingly rivaled by political ones.[26]

LATE NINETEENTH-CENTURY DEVELOPMENTS IN JEWISH SALONIKA

Salonikan Jewish discussions of the possibility of a Greek Salonika only began in earnest after the Greco-Turkish War. The final decades of the nineteenth century, however, were marked by an array of political possibilities and new ideologies. As Greece established itself as a nation and began to actualize the territorial expansionism of the Great Idea, Salonika's Jewish community was undergoing its own reorientation. It had internal as well as external cause to do so. Increasingly, it did not have uniform views on religion, politics, and the best future for the community. The rise of international Jewish organizations such as the Alliance Israélite Universelle and transnational ideologies such as Zionism chipped away at the city's insularity, while political changes within the Ottoman Empire—*tanzimat* (reforms/restructuring, 1839–76) midcentury, and soon after its end, the rise of Turkish nationalism—placed Salonika at the center of Ottoman modernization. Assimilation and secularization also had a significant influence. Greek expansionism was only one factor in an increasingly fluid and complicated social and political landscape.

In the nineteenth century the Ottoman Empire, largely in response to nationalist movements in its Balkan provinces, began its own process of nationalization and modernization.[27] Tanzimat itself was a nationalizing, if not nationalist, movement, which in effect granted Ottoman citizenship to imperial subjects. While tanzimat strengthened the legal powers of Ottoman subjects who had been weakened under Ottoman rule, it had the opposite effect on Ottoman subjects whose legal autonomy had been increased. Some legal arenas—marriage and divorce, for example—remained the bailiwick of religious courts. But civil, criminal, and commerce law all were appropriated by the state. This weakened communal legal authority. In the case of the Jewish *millet* (legally protected religious minority) it also marginalized the community as a whole. As Esther Benbassa and Aron Rodrigue put it, under the new structure community institutions were "basically religious organizations . . . [that] incarnated and strengthened the foundations of ethnic difference on the ideological and symbolic plane, [while] their authority lost the real significance that it had enjoyed in earlier periods."[28] Tanzimat, coupled with the loosening of

Ottoman control over western territories, was incentive for communal reorganization and reconceptualization.

One proposition was Zionism. At the turn of the century, there were more Jews in the region than there were Albanians and Slovenes.[29] But the Jews, alone among the Balkan peoples, had not unanimously embraced a nationalist ideology. Should the Jews too agitate for their own state? If so, where should it be? Would they fare better under such a state than under the Ottomans? Were Jewish interests best served by helping the Ottomans combat the emergent national movements or by developing parallel ones of their own? Such questions had first been posed by Jews living within the Austro-Hungarian Empire, who, like their Ottoman confreres to the southeast, had witnessed the emergence of various nationalist movements, most notably the Magyar uprising of 1848–49. Soon, the ideas that arose in response to them spread to Greece and its surroundings. In the unique setting of Salonika, however, they initially didn't get much traction—certainly not enough to make Zionism seem like a threat to the Ottoman state. As one minister commented to the *Journal de Salonique* in 1911, "We know very well how to defend Macedonia and European Turkey from claimants who are *really* terrible, so why should we fear the Jews?" As for Zionism, "[It] makes us laugh out loud."[30]

Salonika's first Zionist association, Kadimah (forward), was founded in 1899 by David Florentin, editor of the Zionist paper *El Avenir*, and a number of other graduates of the city's Talmud Torah, which had burned down in 1898.[31] Initially, Kadimah was committed to the dissemination of a brand of Zionism that was as much cultural as political, advocating the development of Hebrew as a modern language and a return to Jewish learning.[32] Advocates described it as an expression of pride and unity, and emphasized that it was not an alternative to Ottomanism:

> What do we understand in nationalism? The blood in our veins, inheritance of a long chain of generations who shared memories and feelings, who suffered together—this Jewish blood must pass on clean and complete to our children. This is a principle that we must keep. . . .
>
> We must preserve this sense of unity in our heart. We must not be ashamed of Judaism; [instead] we must develop the sense of pride until we awaken the sense of respect in the hearts of those who try to humiliate us. This sense should live among us forever. And that is the national feeling! . . .
>
> We are good Nationalists and good Patriots—we consider all Jews all over the world as our brothers, as our flesh and blood, and all this does not prevent us from fulfilling our duties to the Ottoman Kingdom, and we confidently say, all Ottoman Jews will always be ready to spill their blood for the motherland. . . . We respect all seri-

ous opinions, but we could not refrain from protesting against all those who try to present the nationalists as anti-patriotic.[33]

So long as this cultural element dominated, the movement met with the approval of the Jewish authorities. Kadimah was allowed use of space in the ḥakham ḥâne (the old rabbinate), and its dissemination of Jewish cultural knowledge and pride was regarded as positive.[34] But when the Ottoman authorities grew suspicious of the possible political implications of Zionism's growth in Salonika, community leaders, eager not to offend their rulers, became more hostile. Much of the Jewish press was anti-Zionist. While *HaMevasser* (Harbinger) was Jewish nationalist, *El Tiempo* and *La Epoca*, both in Ladino, were opposed to Zionism, as was the *Journal de Salonique*.

By 1911, the epithet Zionist had become so dangerous that Jewish Salonika was fractured by accusations—spurious and authentic—that this or the other person was a Zionist. The Turkish newspaper *Hak* (truth) issued reports about a secret Salonikan Zionist society that was agitating in the community, and warned that while Zionism wasn't popular in Salonika, it could become seditious. "All followers of the Zionist idea must know that Jerusalem [i.e., Palestine] is an Ottoman country. If the Zionists want to create an independent government, they should look for a place outside the countries of the Ottoman government. We will be happy to escort them . . . on their trip."[35] To counter such concerns, the leaders of various Salonikan congregations sent a telegram to the Ottoman grand vizier Ibrahim Hakki Pasha in March 1911.

The only ideal of all Ottoman Jews in general, and of Salonika, the cradle of freedom, in particular, is to ensure the happiness and progress of our holy motherland. We are all ready to sacrifice our lives for that cause. . . .

No Jew ever shared—not in the past, not in the present, and will not in the future share—any of the principles of Zionism, which are of no importance. It is evident and known to all that this idea . . . not only does not bring good to the country, but also brings conflict between all Ottoman elements and might create a reactionary movement. In this there is no doubt. Therefore all the members of our community . . . are ready to spill their blood in order to foil this propaganda [that the Jews are Zionists]. Our city committee and our leader, who convened the general assembly, decided unanimously to present your Excellency with these assurances.[36]

As the Balkan Wars loomed, Salonika's Jews made good on this promise. "We swear [our loyalty] to our fellow fifty million Ottomans," declared the community leader Emmanouel Karasso at a huge rally in the Plateia Eleftherias.[37] Yet as before, these overt declarations of support for the Otto-

mans were taken note of in Greece, as the familiar accusation that Jews
were disloyal to Greece became a recurrent theme. As in the Peloponnese
in 1821, Thessaly in 1897, and Epirus in 1913, Greek and Ottoman Jews
alike were caught between being Turkish and being Greek.[38]

The simple fact was, the city's Jews had little reason to be Zionist,
particularly at a time when the mounting Greek movement to "liberate"
Macedonia loomed as their single most pressing concern. As the World
Zionist Organization established its first local branches in Ottoman territories, the Balkan Wars were brewing. Salonikan Jews increasingly fretted
over the impending prospect of political change, and felt that the way to
counter it wasn't by having Jews leave the region but just the opposite: if
as many Jews and Muslims as possible could be brought to Macedonia,
Greece (and Bulgaria) would find it harder to justify their territorial
claims. The constitutionalist reform group Committee of Union and Progress (CUP), angered by the Ottoman loss of much of the Balkans, growing
European intervention in the region, and the long-standing Ottoman political elite, established a Salonikan stronghold in the early twentieth century. In 1908, the Salonikan CUP successfully forced the restoration of
the 1876 constitution; Sultan Abdul Hamid was forced from power and
exiled to Salonika. Among its various efforts to counter the loss of Ottoman control in the Balkans, CUP advocated for the relocation of hundreds
of thousands of Bosnian Muslims to the region (the first families arrived
in late 1910), and urged Balkan Jews to join them too.[39] In a political
context that emphasized the centrality of Jews to Ottoman regional
strength, political Zionism was not much of a contender. While CUP's
success was short-lived, it brought many Ottoman Jews around to a
staunchly pro-Ottoman position. While other nationalist movements,
with their aggressive agendas of nation building, seemed to rule Jews out,
the fuzzy Ottomanism of CUP left room for their existence. As CUP's
support among certain groups gave way to more revolutionary ideologies,
among Jews support for the Ottoman Empire grew. Increasingly, Salonika
was both the geographic and ideological battle line in the Ottoman struggle to maintain imperial integrity—a factor that cannot be overestimated
in its influence on Jewish political orientations in the period.

Political Zionism's lukewarm reception in Salonika stemmed from
other factors as well, many deeply rooted in the city's history, and in
Salonikan Jewry's powerful collective memory of the expulsions from
Sepharad and the Ottoman offer of refuge that came after them. The
long-standing privileges of the Salonikan community, their centuries-
long experience of being the city's majority religious group, and their
great economic and cultural success all contributed to the sense that
Salonika, despite its troubles, was home. Jewish leaders from the outside
were struck by this loyalty to topos. Theodor Herzl's famous visit to

Athens in 1898 gave some of Salonika's community leaders the chance to meet directly with Zionism's founder, but didn't do much to boost the movement's popularity there.[40] When David Ben-Gurion paid a visit, the city's palpable anti-Zionism made him uncomfortable. Indeed, he could scarcely recognize its Jews, so astonished was he by their physicality and the sight of what he called "a Hebrew labor town, the only one in the world."[41]

The absolute centrality of Salonika to the identity of its Jews is reflected in the fact that since the sixteenth century, Salonika had itself been regarded as a sort of "Zion," a promised land of refuge, the "Jerusalem of the Balkans."[42] More than three centuries later, in the wake of the Holocaust, the surviving remnant of its Jewish community still remembered it by the same name. In comparison to the real thing—an economically underdeveloped Ottoman outpost populated with Muslim Arabs—Salonika looked like an improvement. As some put it, "What is this Palestine you're telling us about now? This is Palestine."[43] This powerful locative sense, above all, was the sentiment most responsible for Jewish Salonika's lack of interest in movements that might take its members away from their city. To be a Salonikan Jew was not simply to be Jewish; it was to be Salonikan.

But Zionism was neither the first transnational Jewish movement to take root in the Balkans nor the most successful. Far greater influence was commanded by the Alliance Israélite Universelle, formed in France in 1860 by Adolphe Cremieux, known best as a forceful advocate of the 1848 Revolution. The alliance was formed partly in response to the infamous Damascus affair of 1840, one of the best-known blood libels of the nineteenth century, and its schools aimed to further Jewish equality through education and foster Jews' integration into broader society. To some extent what this necessarily entailed was secularization. Alliance schools in the Balkans, as elsewhere, created a new generation of semisecularized, Francophone (and often Francophile) youths. Many hoped to travel to the West to pursue professional careers. For the first time in centuries, members of Salonika's long-closed Sephardic community began to look to the outside world as the place of future opportunities. Photographs of the period show that this new generation began to cast off the traditional markers of community membership. They exchanged their parents' clothes for Western styles, became looser in their religious observance, and mingled with their non-Jewish neighbors. Cafés and movie theaters sprung up around Salonika's shoreside promenade to cater to the new cohort of semi-Europeanized Jewish bourgeoisie.[44]

Between the early 1860s and the start of the First World War, the alliance established a modern educational network throughout Greece, the Balkans, and the Levant; in 1873, the first alliance school came to Salonika, and ultimately the city had nine.[45] Its alumni association (Association d'anciens élèves de l'Alliance Israélite Universelle), founded in 1897, emerged informally as the major rival organization to the Zionist Kadimah, although there was a good deal of cross-fertilization between the two as well.

That the alliance had multiple and mixed effects on the Jews of the region is attested to by the various international publications that commented on its activities. The *African Times and Orient Review*, a black-run paper with an editorial commitment to the assertion of all minority rights against the oppressions of the European establishment, lamented the alliance's royalist, secularizing tendencies and decidedly bourgeois agenda. A 1912 editorial captures quite well the debate within Judaism over its activities:

> Though the Alliance Israélite has done a great deal of good in bringing light into the dark places of the [Ottoman] East—and the education imparted in its schools is indeed excellent—it has a great deal to atone for. By weakening the allegiance which their co-religionists owe to their ancestral faith, the Alliance, by the evil example it sets both at home and abroad, not only incurs the opprobrium of the more orthodox section of the Jewish community, but is guilty of creating a type of person who is a gross material opportunist, neither good Jew nor good patriot.[46]

The alliance created a new cadre of secularized, Westernized Jews who to some degree turned away from old traditions. It was the alliance's first generation of graduates who confronted the prospect of political change in Salonika, and to an extent, it was the alliance that helped ease the transition, as alliance graduates were likely to find the new circumstances less jarring than would the traditionalists.

This time, however, social circumstances were already primed for change. The Young Turk revolution of 1908 had brought with it the language of civil liberties, and Zionism, after its first relatively unsuccessful advent in Salonika, had gained some ground. The Bene Zion and Benot Zion (Sons and Daughters of Zion) associations, founded in the wake of the Zionist leader Ze'ev Jabotinsky's 1908 visit to the city, were viable, if not booming, organizations; the new chief rabbi, Yaakov Meir, installed in 1907, was a vocal supporter of Zionism.[47] Other forms of political activity had also developed. Jews were organized against the anti-Semitic campaigns of the Greek press. Many members of the working class had joined the Socialist Federation. A popular sporting association, Maccabi, also founded in 1908, had grown over the years to have a membership totaling in the hundreds.[48]

From Venizélos to the Balkan Wars

Recasting what it meant to be Jewish became more pressing as Greece recovered with astonishing alacrity from its 1897 defeat, gaining a new political star in the process: Eleuthérios Venizélos, who commanded huge respect in Greece, and was accepted in Europe with a respect and admiration unknown by Greek politicians since Ioannis Kapodistrias. A Cretan assemblyperson who was an ardent and active advocate for énosis, Venizélos became prominent in the government established after Crete was granted autonomy as part of the 1897 postwar settlement. In 1905, he resigned and then led a revolt that declared the unification of Crete with Greece. Greece's Prince George, awkwardly stuck between the great powers, who didn't support rapid unification, and Venizélos, ultimately had to step down.[49] The twofold legacy of Venizélos's rise to power was a long-standing and mutual enmity between Venizelists and royalists, and a persistent push for further territorial expansion. Venizélos's party, the Kómma ton Filefthéron (Liberal Party), was to dominate Greek politics for more than two decades.

The post–Greco-Turkish War period was extremely turbulent for Greece. In the twelve years between 1897 and the Goudi coup of 1909, which brought Venizélos to power, Greece's two main political parties virtually collapsed.[50] There were repeated outbursts of urban riots (most notably in 1901 and 1902).[51] The monarchy was marginalized, and parliamentary procedure broke down.[52] In the midst of this turmoil, the one platform that remained consistent was for the "liberation" of Macedonia. The Venizelist agenda was twofold: to protect Greeks outside of Greece from the anti-Hellenizing efforts of the Young Turks, and to Hellenize the territories within Greece that had sizable minority populations.[53] Both goals, in different ways, bore extremely negative implications for the Jewish communities of the region.

As the struggle for Macedonia intensified Greece modernized its military in preparation. The long-standing guerrilla conflict in the region continued, as did the ecclesiastical jockeying for dominance between the Bulgarian and Greek Orthodox churches. In 1912 the Balkan Wars broke out. Incongruously, Greece at first joined with Bulgaria, Serbia, and Montenegro against the Ottomans. Much to the surprise of Russia and the West, the Balkan alliance soundly defeated the Turkish forces. But the conflict turned promptly into a competition between the former allies, who rushed to beat one another to take possession of the territory abandoned by the routed Ottomans. In October 1912, the Greeks beat the Bulgarians to the punch. The highly educated, urban, elite Jews of Salonika suddenly found themselves under Greek occupation, soon to be ruled by "the political elite of [a] Christian peasant societ[y]."[54]

From the Greek point of view, the conquest of the city had divine timing: it fell on the feast day of the city's patron saint, Demetrios. The "liberation" of Salonika, like the Greek War of Independence ninety years earlier (commemorated on March 25, the Day of the Annunciation), coincided with a Christian religious narrative, a Greek Orthodox teleology according to which Saint Demetrios, protector of the city, would personally free the Greek Christians from Turkish rule. The annual festival of the city's conquest is a commemorative event of simultaneously nationalistic, civic, and religious import. The precise timing of the liberation, in bringing contemporary history into line with Greek Orthodox tradition, didn't simply bring Salonika across the border from Ottoman to Greek territory. It also rewrote the city's Jewish past. The conquest of Salonika expanded Greece's frontiers, but also symbolically underscored the Christian underpinnings of Greek nationalism. The conquest was a *Christian* conquest, Saint Demetrios's vindication, its religio-apocalyptic overtones an adumbration of the future Christianization of its population.

Greece's acquisition of Salonika rendered it, overnight, home to the largest Jewish city in Europe, again, in terms of the percentage of the population. And it rendered Salonika's Jews the primary target of and obstacle to Greece's Hellenization of the region. The three brief decades between the Balkan Wars and World War II were the high-water mark of Greek Jewish history. But they also were the most turbulent decades in the history of the one community—Salonika—that has come, more than any other, to be coterminous with the category Greek Jewry. The story of the Hellenization of Salonika is arguably the final chapter in Greece's bumpy transition from imperial territory to independent nation—one that went on longer than the Ottoman Empire itself.

Chapter 5

BECOMING GREEK: SALONIKA, 1912–23

The abolition of all official recognition of the Jewish Community and of their right to manage their internal affairs, especially as regards the "état civil" of the individual, the imposition of Greek, a foreign and unknown language, in which all books and registers must be kept, and such-like measures, combined to make the Jews (and not the Jews only) long for the return of the Turks; or, better still, the creation of an autonomous Macedonia under the protection of "Europe" or some one European power.[1]

IN 1912, Salonika far surpassed what today would be called a multicultural city. With no one official language, close to half its 160,000-strong population was made up of Sephardic Jews, with the remaining half divided between Orthodox Christians, Muslims, *Dönme* (Jewish converts to Islam), Armenians, Slavs, Roma, and foreign nationals.[2] Dual nationality was commonplace, with up to a quarter of the city's residents claiming legal status under more than one political regime. As a Western visitor wrote in 1918:

Imagine a town where the languages commonly and regularly spoken are old Spanish, much adulterated, Greek, Turkish, Italian, Bulgarian, Serb, Roumanian, and French; where every one has changed his subjection at least once during the last five years,—from Turkish to Greek,—and where before that several thousands of people had all sorts of claims to European nationalities . . . (under which one brother in the same family would be "French," another "English," another "Italian"), perhaps without one of them being able to speak a single sentence in the tongue of the nationality he claimed.[3]

But while this sort of categorical confusion was not at odds with Ottoman rule, it did not make sense within the context of a new, ethnically homogeneous nation-state. The Jewish leadership of Salonika recognized, as the president of the Jewish Board of Deputies put it, that "many members of the Jewish race and religion [were] confronted by the necessity of transferring their allegiance from one state to another." This in and of itself was not a concern. At issue was whether the transition would "entail a limitation of their rights and privileges as free citizens," and whether the

Greek government would be willing to support "the principle of religious equality and liberty."[4] From the start, the transition to Greek rule was understood as one that would likely place Greek Orthodox Christians above the members of other ethnic groups. Salonika's Jews, long the de facto, if not literal, rulers of the city, stood to lose a good deal. While they long had "virtually governed themselves in accordance with their own customs and traditions," with "no sort of assimilation . . . expected of them," now quite the opposite would be the case.[5]

ZION IN THE BALKANS?

With the arrival of Greek troops, a sense of dismay settled over the city's Jews. As one observer drolly commented, "There can . . . be no doubt that the Jewish community of Salonika looks forward to a perpetuation of the Greek régime with a total absence of enthusiasm, if not with serious apprehension."[6]

During the war, the Jews had done what they could to resist the Greeks. Community representatives had told Kamil Pasha, the Ottoman commander, that Salonika's Jewish banks would give generous financial assistance to the Turks if they would not give up the fight against the Balkan alliance.[7] Once Greek forces occupied the city in 1912, the Sephardim were part of the jockeying between the great powers, Greece, and Bulgaria to determine Salonika's ultimate status. Even after the arrival of the Greek troops, Salonika's Jews found it inconceivable that their city would be Greek. Hastily, a proposal was put forward for the internationalization of the city.

Secret correspondence between the Salonikan Jewish leadership and international Jewish organizations presents the majority position of the city's Jews. On December 15, 1912, David Florentin, the editor of *El Avenir* and deputy chair of *Maccabi*, wrote a "top secret" letter to the president of the Zionist Organization Committee in Berlin. "At this time, when the Great Powers are preparing once again to divide the Balkans, there is an acute necessity to warn the Jewish communities in Europe of the dangers this presents to the important Jewish community of Salonika." If the Greeks took possession of the city, "the port, which has been serving as a gateway for the entire region, will be cut off from the surrounding area." Salonika's economic vitality would end. "We cannot see any logical reason for a development in public wealth. All the efforts of the government will no doubt be directed toward one end: the Hellenization of the city." Florentin grimly predicted that "the annexation of Salonika by Greece means the end of the economic prosperity of the city's

Jews. Any other solution will be better for Salonika." To be part of Bulgaria or a free Macedonia was preferable to unification with Greece. But best of all would be total independence.

> International Salonika, like Tangiers or Dalien in Manchuria, shall be built on economic prosperity, through the encouragement of different countries, because it will cease to be the arena of endless quarrels and stratagems. There are chances that it will continue to be a resource for most of the Balkan countries. Internationalization will be, according to all signs, one of the best solutions for us, the Jews.[8]

The plan to internationalize Salonika, though popular with Jews in the immediate region, was rejected by international Jewish groups, which argued that its success would undermine Zionism. Other religious establishments in the city, notably the Greek Orthodox and the Dönme—who jointly wrote to Venizélos expressing their support—were advocates.[9] But Western Jewish leaders contended that the goal of a Jewish state had greater prospects in the absence of any obvious Jewish safe house. If the famously Jewish Salonika gained independence and autonomy, the claim that the Jews needed their own homeland would look less urgent and lose credence.[10] Just years after its invention, Zionism was already a movement paradoxically dependent on the misfortunes of Jews as much as responsive to them. Some Salonikan Jews, already for the most part opposed to Zionism, were bitter about the role it played in this penultimate attempt to free their city. Bulgaria was also bitterly opposed to the idea of internationalization and threatened an economic blockade of Salonika. Since one of the key assertions in the plan's favor was that it would maintain the city's links to the broader region, this threat doomed it to defeat. In the long run, the plan damaged Salonika's Jews, whose attempt to break free from Greece was regarded by many Greek Christians as another instance of Jewish disloyalty.[11]

The defeat of the internationalization plan gave way to an interesting, if short-lived, Zionist hybrid, and resulted in a brief and exceptional period of ideological unity among Salonika's Jews. The Jewish leadership of the city—assimilationist and Zionist alike—came forward with a new scheme: to create a politically neutral buffer state, with Salonika as its capital, that would ease tensions between Greece, Bulgaria, and Serbia even as it allowed all three collectively to dominate it.[12] This expression of "territorial Zionism"—the advocacy of the creation of a Jewish national home, but outside of Palestine—brought together multiple Salonikan factions, Jewish and not, and had the muted support of some members of the Venizélos government.[13] Again, the Central Zionist Organization demurred to intervene on the Salonikans' behalf—not surprisingly. Rena

Molho summarizes the situation succinctly, noting "the paradox of asking the Central Zionist Organization to intervene on behalf of the Jews of Thessaloniki, and help them not to be the first to emigrate to Palestine, but to remain a city of the diaspora."[14]

The city was occupied by Greek forces from 1912 until the resolution of its status almost two years later. Though Venizélos briefly entertained the various proposals for Salonika's independence, this is probably because he viewed them as stepping-stones on the path to Greek ownership of the city; at least they kept it out of Bulgaria's clutches. But there was no question in his mind that Salonika was rightfully Greek. Before the second Balkan War he had declared, "Salonika belongs to Greece by right of history and by right of conquest. What I can tell you is that Greece is ready for any other sacrifice for the maintenance of the Balkan Alliance [of the first Balkan War]. But to give up Salonika,—that will never be done,—never, never."[15]

Makedonía, a fiercely nationalist newspaper that soon became the primary media outlet for extreme anti-Semitic views, set the tone for the new Greek Salonika with its inaugural issue of October 28, 1912. The front page was given over to the Greek flag and the headline "*Zíto i Eleftheria*" (Long Live Freedom). *Embrós* (Forward) and other Greek dailies looked much the same.[16] For the city's Jews, however, freedom would prove a relative concept.

Under the terms of the Treaty of Bucharest (August 10, 1913), Salonika and its hinterlands formally became part of Greece. While the treaty made special provisions for the Koutzo-Vlach inhabitants of the region, no special mention was made of the Jews.[17] Soon after, Salonika's chief rabbi sought an audience with King Constantine, who had come to the throne after his father, King George, was assassinated in March 1913. The king was well aware of the economic importance of his new Jewish subjects. With disarming honesty, the rabbi flatly declared that he hadn't wanted the Greeks to win. But since they had, he was ready to swear his community's loyalty to the Greek king:

> We tried our best to support the course of Turkish domination in Macedonia, and we Jews would have been willing to sacrifice ourselves to preserve that Turkish domination, should it have been possible. I must report in all candor that I would have taken up arms if that had not been an impossibility, in order to prevent the fate which befell the Turks. We have now adjusted to the realities brought upon us by Greek rule and domination.[18]

The Jews would be loyal to Greece, he emphasized. The community was aware of its responsibilities to its new government. Constantine, in turn, assured him of the Jewish community's ongoing economic security and

Greek goodwill toward the Jews. The rabbi reported the conversation to his constituents, but for the most part they remained pessimistic about their new status as Greeks. In a follow-up communiqué, the king promised that Greece's Jews would never be a target of hostility because of their religion; nor would Greek courts ever entertain accusations of blood libel brought against them.[19] Greece's readiness to reassure its new Jewish citizens was largely a response to the fact that Germany, Spain, and Austria had all offered them citizenship and "protection."[20] The Greeks were eager to show the world that no citizens of theirs needed outside guarantees of security.[21]

Salonika's Jews hoped, in effect, that the Greek state would prove willing to reestablish many of the legal protections that Jews had enjoyed under Ottoman rule. Specifically, the Jewish community requested official state recognition of the grand rabbi and the Jewish Communal Council; the reestablishment of religious law halacha and courts (*bet din*) for the adjudication of inheritance, marriage, divorce, and other civil matters; the right to levy taxes on the Jewish community for the maintenance of Jewish schools and institutions; and most critically, proportional representation in Greek institutions—on the city's municipal council as well as in the courts and gendarmerie.[22] The same request would be put to the government at the close of the Great War, when the drawn-out settlement proceedings presented themselves to Jews as a chance to try to gain terms that were not won at the end of the Balkan Wars. What was at stake, in essence, was the effort to retain Ottoman imperial communal legislation within the context of the Greek nation-state. This makes adjudication of the many accusations made in the period of systematic, state-sponsored "anti-Semitism" on the part of the Greek government difficult: anti-Jewish bias in Salonika was framed not simply with the language of universal human rights and equality but also with the imperial language of privileges and communal civil autonomy—much as had been the case on Corfu after the cession of the Ionian Islands to the British in the mid-nineteenth century. The perpetuation of millet-based religious categorizations was a process in which Orthodox Christians and Greek Jews alike participated, although for quite different reasons.

In the decade following the Treaty of Bucharest, Salonika would undergo more physical and cultural changes than it had over the course of the previous centuries. The Jewish community was powerfully affected by external pressures and also dramatically reshaped itself from within. Zionism, earlier scorned by much of the community, had a modest efflorescence; socialism, long an alternative ideology embraced by the Jewish working classes, became joined to Zionism; and assimilationism, in light of increasing Greek cultural dominance in Macedonia, ceased to be a vague alternative position, and became instead a matter of law and necessity.

Greece in World War I

Look out and see what was the Greece you handed
over to the Liberal party and what is the Greece it deliv-
ers to you; be careful that you do not deliver her to
your successors smaller.
                    —Venizélos to the Greek House
                    of Representatives, September 28, 1915

From the start of the Great War, two Greek policies toward involvement
emerged. Each coalesced around a popular figure. King Constantine I supported a policy of neutrality, while Venizélos, the leader of the Liberal
Party and prime minister of Greece, advocated for intervention on the
side of the entente.[23] Their opposition ultimately led to a constitutional
crisis and deeply divided Greek society. The origins of the so-called interwar national schism (*ethnikós dihasmós*) lie in Venizélos's and Constantine I's utter inability to come to a compromise. The clash between
them defined Greek politics until the Second World War and beyond. Salonika's Jews would have to negotiate their newfound Greekness against
the backdrop of a bitterly divided nation. What would have proved a
monumental task even under the calmest of circumstances was all but
impossible in the context of war.

King Constantine's neutrality was really a pro-German bias; in May
1916, he turned critical Macedonian sites over to the Germans and Bulgarians.[24] The entente, increasingly impatient with what it perceived as
the monarchy's duplicity, became strong advocates of Venizélos. At the
war's end, Viscount Northcliffe described the entente's contempt for the
Greek monarchy: "Greece, or, to speak correctly, the King of Greece and
his pro-German court . . . misled and hoodwinked the chancelleries of the
Entente. . . . They overthrew the Greek Constitution, and imprisoned or
drove into exile the statesmen who really represented the Greek people."[25]
For his part, Venizélos saw the First World War as an extension of the
Balkan Wars; if German power in the Balkans was not quickly checked,
the Greeks "[ran] the risk . . . of seeing the great Greek achievement which
was accomplished by the wars of 1912 and 1913 becoming of a very short
duration." The war had come too quickly after Greek expansion into
Macedonia, he argued, "before Greece had the time to organize her new
possessions."[26] To opt out of the war would be tantamount to relin-
quishing the newly taken territory. The possibility that war offered for a
push into Asia Minor was also on his mind, and Venizélos framed support
for the Allies as a defense against "the destruction of the rest of the Hellenic Race, the uprooting of the Greek race in Turkey."[27] In the consum-
mate expression of the Great Idea, Venizélos explained that the First

World War would provide, at long last, the chance "to bring about the creation of a great and powerful Greece, constituting not an extension of the State by conquest, but a natural return to the limits within which Hellenism has flourished ever since the prehistoric period."[28] The Venizelist/royalist dispute ultimately did untold damage to Greece, such that by the middle of World War I, Greece was close to a state of civil war.[29]

To the European troops who were sent to the Balkans to fight in the war, it was clear that their leaders didn't know what they were getting into. British troops arrived in Salonika in 1915. They soon found out that Macedonia, in both its terrain and politics, provided quite a challenge. As one observer wrote,

> Most of us . . . who have gained an intimate knowledge of Macedonia are willing to hazard a guess that the authorities who sent us there knew very little about the country, had no idea of the damage that malaria could do to Western troops, and did not dream that the land would soak up man-power as a sponge soaks up water and give so little in return.[30]

By the war's end, the conflict between the royalist and Venizelist camps resolved, albeit unsteadily, in favor of the latter. Venizélos emerged as the darling of the great powers, and while the Liberal Party and the monarchy would remain enemies, the Venizelist vision—with the Great Idea at its heart—prevailed. Greek politics, however, were torn apart as never before. The parliamentary debates of 1917 over the war, among the fiercest in Greek history, set the tone for decades.[31] Great power meddling in Greek affairs would also continue, heedless of the implications of the region's internal divisions. But at the end of the war, the enthusiasm with which the West had welcomed Venizélos would rapidly give way to a lack of faith in Greece's ability to function as a sovereign power. The instability bred by the great powers' interventions in the Balkans and particularly Greece would remain a dominant theme throughout the twentieth century.

SALONIKA DURING THE WAR AND THE GREAT FIRE OF 1917

The arrival of Allied troops in Salonika in October 1915 blocked Germany's attempt to gain full control of the Balkan Peninsula, maintained the sovereignty of Serbia, and was designed to guarantee British regional dominance—critical, given that, as the Allied forces' official war correspondent in the Balkans put it, "Egypt and the Suez Canal have lost much of their importance as the gatehouse of the East now that the trans-Balkan railway runs straight through from Berlin to Baghdad."[32]

The drama of the Balkan Wars was thus rapidly followed by the even greater disruption of the Great War, which brought the Allied forces to Greece's doorstep, and highlighted the extent to which Salonika was seen as "a practical guarantee [of] the great and vital interests which the Allies . . . possess[ed] in the Balkans."[33] German bombardments inflicted losses on the Allied barracks and the civilian population. A major turning point came in 1917, with Venizélos's restoration to power and his declaration that Greece was now on the side of the Allies.

Economically, the war was a boon to the Salonikans, but of a frenetic, confused sort. The Allied war correspondent wrote that the army's presence was "exceedingly profitable to them," adding that "in the term Greeks, I include the large Hebrew population of Salonika, which is of Greek nationality." Martial law had been declared in the city in June 1916, followed by the so-called Salonika Revolution of 1916, the arrival of Venizélos's rump government that fall, and a year later, the expulsion of King Constantine. On establishing his breakaway government in Salonika in October 1916, Venizélos described Greece's political situation as one verging on civil war: "We consider Greece," he said, "to be a kingdom with two Governments in it, as in the case of all countries at civil war, although actual civil war is the development we are trying to avoid."[34]

The military and political developments brought huge business to Salonika. The arrival of the Allies, in particular, "was the sort of opportunity for money-making that the local Greek and Israelite population could not have surpassed in their wildest dreams." Regional trade had steadily declined since 1912, the result of the Balkan Wars, a cholera epidemic, and a generalized economic crisis. By some accounts, over five thousand merchants had lost everything. In response to the crisis, the Turkish government offered asylum to affected Jews, including in the offer a seven-year tax exemption. As a result, about two hundred Jews moved to Constantinople.[35]

The working classes were also suffering. Labor competition at the port, in particular, was vicious. Already by the mid-nineteenth century, not all those who wanted work there could find it, and Jewish porters and other dockworkers fought against one another to gain the upper hand. When the artist Edward Lear visited in September 1848, he was taken aback by their assertiveness.

> The enthusiastic Israelit[e porters] rushed into the water and, seizing my arms and legs, bore me out of the boat and up a narrow board with the most unsatisfactory zeal . . . and finally throwing themselves on my luggage each portion of it was claimed by ten or twelve frenzied agitators who pulled this way and that. . . . [The] scene changed in a few minutes to a real fight, and the whole com-

munity fell to the most furious hair-pulling, turban-clenching, and robe-tearing, till the luggage was forgotten and all the party was involved in one terrific combat.[36]

When Greek porters and stevedores grew in number and lobbying power after 1912, an already competitive situation became brutal. "When Greek ships [came], which arrived daily in Salonika's port (and the number of Greek ships was increasing), the ship owners would ask immediately if the workers [who met them] were Greeks or Jews, and they refused to hire Jews. Thus Jews were forced to work only with European ships."[37] In response, Yitzhak Ben-Tzvi, the future president of Israel, traveled to Salonika to invite Jewish stevedores to come work in the Jaffa port; Ben-Tzvi wanted to replace Jaffa's Arab port laborers with Jews. About twenty took him up on the offer. Just as Greeks worked to establish dominance in the Salonika port, importing Greek stevedores to that end, Zionists, with similar motivation—to help establish Jewish control in Palestine's ports—imported Jewish stevedores. In a parallel to the tension between Jews and Greeks in prewar Salonika, by the 1930s Arabs in Palestine were lodging complaints about the establishment of a Jewish-run port in Tel Aviv that was rapidly eclipsing Arab Jaffa. In the 1920s and 1930s, Jewish ship workers and stevedores were in especially hot demand in Palestine, much as Greek ones were in Salonika.[38]

The arrival of the war now revived the economy and then some. British diplomatic documents suggest that at first, Salonika's Jews welcomed the war both because of the brief economic boon it brought to the city and because it presented the chance of a political reshuffle.[39] Prices were hugely overinflated, and the profit margins were enormous. Rental properties were offered at exorbitant prices. One, a poorly built villa with dubious sanitation facilities and no furniture, was rented out for more than £130 per month. Another fetched £200 per month, a rate twelve times its price under nonwar conditions.[40] Under these circumstances many of Salonika's residents grew richer, most dramatically those who were well-to-do to begin with. For the poor, though, the situation brought great hardships. One eyewitness, Yitzhak Immanuel, recounts that "prices went up, and produce disappeared from the market to such an extent that it was very difficult even to buy bread. Many Allied soldiers sold their bread [to civilians] for twelve francs per kilo."[41]

There were also some happy interactions between Allied troops and the local population. Chief Rabbi Yaakov Meir asked the Allied commanders to give Jewish soldiers time off at the Jewish holidays, and "almost every Jew fulfilled the *mitzfa* [sic] of hospitality on the holidays with happiness." Jewish soldiers went to civilian homes for the Passover seder, and the chief rabbinate organized a huge public seder in the courtyard of the Talmud Torah for all soldiers who didn't go to a private home for the meal.[42]

The orgy of moneymaking and entrepreneurship was brought to an abrupt and brutal halt by the huge fire of 1917.[43] Within the course of two days, August 5–6, it destroyed up to three-quarters of the town within the city walls, including the business district and the waterfront.[44] The fire instantly created sixty thousand refugees.[45] The Allied forces, who had been living mostly beyond the city walls, were largely unaffected. Just as it was joining the Allied campaign, the Greek government suddenly had a major domestic disaster on its hands. The less charitable among the Allied witnesses viewed the fire as "a judgment upon the greed of Salonika," a punishment for the extortionist economy the war had produced.[46] As for the Jews, who used Hebrew letters in their dating system, it was not lost on them that 1917—*tav, resh, ayn, zayn*—could be taken to spell the word *tav'erah*, or "great burning." Thus, 1917 was remembered as "the year of the burning," and was commemorated in countless poems and memoirs as an apocalyptic as well as worldly event.

> 30th of *Av*: Lamentation on the City
> The community of Saloniki, glorious city
> The day you were burned, the day you were breached
> And inside your heart the flames spread
> Your rupture is great, your wound is wide
> Because the poor and the miserable are without number
> Only this shall be your comfort:
> Universal peace and a return to Zion.
> —Anonymous Salonikan poem[47]

יום שלשים אב בכי על קרת
עדת שלוניקי, עיר הפאארת
יום נשרף נפרץ פרץ
בתוך לבך פשתה הבערת
שברך גדול רחב פצעך
כי אין מספד לדל ולאביון
רק זאת תהי נחמתך:
שלום תבל ושיבת ציון

Salonika was vulnerable to fire. Its houses were of wood and built cheek by jowl beside one another along narrow streets. Its firefighting equipment was pathetically inadequate. The *Balkan News* noted that the fire brigade's hoses were so riddled with holes that more water leaked out of them than came out of the nozzles; "it would [have been] more advantageous to lay the pipe sideways on the fire."[48] Late summer was the driest and hottest time of the year; great strains had been placed on the city's water supply by the influx of thousands of Allied troops and the Venizelist governmental apparatus. Allied commanders were hesitant to use the city's water reserves to fight the conflagration, deploying instead a strategy of blowing up buildings in the fire's path so as to prevent it from spreading.[49] "The Jews were forced [both] to escape the fire and flee the bombs."[50]

Salonika was no stranger to fire. In early September 1890, a large one, likely the work of arsonists, had destroyed swaths of the city—twenty hectares.[51] In 1874, another conflagration (known in the city as "the burning of the Francis family") had struck in the middle of the night of February 12, when anonymous persons set fire to the four corners of the new home of the wealthy Jewish Francis family. While it didn't destroy other parts of the city, the event was a huge shock to the entire populace. Six thousand people, including the Greek metropolitan, attended the family's funeral, and even the *New York Times* had carried a notice.[52]

The 1917 fire was to be the most famous and destructive of all. It struck in the same area as the fire of 1890, but with even greater savagery. Its aftermath would be different as well. In the time elapsed since 1890, the city had become part of Greece; with the destruction of large portions of its most desirable real estate, the Greeks now had the chance to play a decisive role in rebuilding. While the city's earlier architecture and layout had marked it as "Oriental" and multiethnic, the new Salonika that emerged after the 1917 fire was Hellenic and decidedly European. Grand stone buildings and wide avenues would replace the wooden houses and narrow streets; churches would take the place of minarets and the arched doorways of synagogues.

After the fire, hastily imposed Greek legislation forbade reconstruction undertaken on an individual basis, laying the ground for a systematic program of urban redevelopment.[53] The city's Jews in particular saw this policy—to a degree quite accurately—as designed in large part to limit their future influence.[54] A royal decree issued on September 14 turned reconstruction over to the Salonikan authorities, and a plan was drawn up by a group of civil engineers and architects made up exclusively of Greek Orthodox and Western Europeans. Subsequent legislation gave the government wide latitude to redistrict burned areas, set the value of lost property, and buy out what land it wished.[55] Newly districted property was to be sold out in parcels by the government; the former owners had first claim to them, but the selling price would be fixed. Huge demonstrations ensued—eight thousand Jews marched to protest the law, which in effect gave the Greek government the right to confiscate land. In reaction, the Venizélos government promised to turn the profits from land sales over to the original owners, but there is scant evidence that this was ever done.[56]

Angry victims argued that the fire was set on purpose, with anti-Semitic intent.[57] There is no evidence for this theory. From the point of view of the devastated victims, the Greek legislative response to the fire was ruthless, but it was in keeping with the international urban modernization policies of the period. Law 1394, for example, used as its template legislation passed following the San Francisco fire and earthquake of 1906.[58] Various accounts, some with a conspiracy theory edge, and others picturesque, circulated. One had it that a Greek widow frying eggplant had overturned the cooking pot by mistake, and the hot oil had caught on fire and quickly gone out of control.[59]

Whatever its genesis, the fire spread quickly. By four in the afternoon, an hour and a half after ignition, forty homes were gone.[60] By eight in the evening, half the city was on fire. The whole area was one great incendiary device; wooden houses and storerooms filled with oil, alcohol, and spirits kept it burning for hours.[61] The scene was one of utter chaos, confusion, and even macabre beauty. One reporter wrote of the surreal juxtaposition of the fire and frantic residents fleeing with their possessions on their backs with the unexpected sound of Italian music pouring out of a home in a neighboring, unaffected quarter. By the next day, though, there was no music anywhere. "One can say that all of Salonika is in flames. The spectacle is terrifying."[62] Only a rapid change in the wind's direction prevented the fire from destroying the entire city.

The fire devastated Jewish businesses, civic organizations, and neighborhoods. The very institutions that under normal circumstances would come to the assistance of Jews in times of calamity were themselves rendered useless.[63] Between forty and fifty thousand Jews were displaced by the fire, along with ten thousand Muslims and between ten and fifteen thousand Christians. Eight thousand buildings were destroyed, spread over 120 hectares of land.[64] On top of the material losses, the Jewish community suffered immeasurable cultural, religious, and intellectual devastation.[65] One of the largest Jewish libraries in the world was destroyed, along with nine other rabbinical libraries; six hundred torah scrolls, some very ancient; eight schools, five of them alliance; a rabbinic seminary; and all the buildings and archives of the chief rabbinate.[66] The publishing house Etz Hayyim, which issued all the city's scholarly books, was razed.[67] Thirty-two synagogues were consumed in the fire, and the headquarters of the associations Intimes, Nouveau Club, and Kadimah were gutted.[68]

Tents and improvised barracks were set up to house survivors, and the Alliance Israélite, which had some intact properties, began the distribution of food, blankets, and clothing. With the legislation that prohibited rebuilding, the miserable ad hoc living conditions soon began to look permanent, and many Jews started for the first time to think that it might be best to leave altogether.[69] During the brutal winter of 1918–19, 1,569 Jews died within one month. In 1919 even the chief rabbi left the city.[70] The legislation on reconstruction continued overwhelmingly to be opposed by Jews, who had suffered the gravest losses in the fire and now stood to lose the most in the reconstruction scheme. Most of the fortunes lost had not been insured; what had been was compensated at a rate of only 60 percent.[71] The Athens government acknowledged that it was more aggressive in the assistance extended to Greek Christians than Jews, but argued simply that while the Jews could rely on a huge international Jewish philanthropic network for assistance, Greek Christians could expect no equivalent level of international, outside interest in their plight.[72]

While some Jews in the city were Venizelist and supported the reconstruction plan, for many others, long suspicious of Venizélos, the fire's aftermath was taken as grim confirmation that the New Greece would be a most inhospitable home.[73] As had long been the case with controversial topics, all aspects of the fire—its inception, claims of losses, and the legislation for reconstruction—were fiercely debated in Salonika's many newspapers.[74] While L'Opinion and El Liberal argued a Venizelist position, most ardently protested the Greek policies.[75] For the following three years various lawmakers, Jewish community leaders, international Jewish organizations, and the general citizenry argued. In the 1840s, numerous Athens residents had fallen victim to the government's grand plan of creating a new, neoclassic city that would match the grandeur of antiquity.[76] Many Salonikans now had to suffer their city's turn at architectural Hellenization and Europeanization.

What was rebuilt, however, was not simply the physical city. Its human texture was completely reworked too. The Jewish community became more overtly politicized and dichotomized, divided between Zionism, socialism, and—less markedly—Venizelism. The Greek socialist party, which in 1918 formulated its own proposal for a solution to the Eastern Question, appealed to many Jews for its embrace of the working classes, and its explicit belief that class trumped race and nationality.[77] Some saw socialism as a solution both to class problems and the "Jewish Question." Avraham Ben-Arroya, a Sephardic Jew from Philippopolis who came to Salonika in 1908 to participate in what he called the "revolutionary ferment," recalls, "I went to Salonika [because] I wanted to prove empirically [be-Ofen Ma'asi] that the masses of Jewish workers [were] not outside the laws of the socialist doctrine, and that it is also relevant to them. I was sure that the social solution would bring within its wings also the solution to the Jewish national problem."[78] Ladino socialist newspapers such as Avanti, Il Journal de los Laboradores, and La Solidarita Obradera appeared as early as 1912.

The forces of secularization and Westernization continued apace after
the war. Emigration from Salonika to Athens, and particularly to the
Americas, quickened. But while Zionism grew in popularity as an ideology, few Jews left for Palestine.

A tumultuous series of events—the 1897 Greco-Turkish War, the 1912
Greek conquest of Salonika, the arrival of the Allied troops in 1915 and
the Venizelist government two years later, the great fire of 1917, and finally the rebuilding of the early 1920s—amounted to an ineluctable
tide in the direction of Hellenization. Arguably, any one of these alone
would not have been enough to tip the balance. Taken together, they
were a series of body blows that came so quickly one on the other that
the Jews of Salonika never fully regained their footing. And the final and
most devastating in this series of events was soon on the horizon: with
the Asia Minor disaster and the population exchanges of 1923, Salonika
would be set to become a definitively modern, Hellenic, and Orthodox
Christian city.

THE *KATASTROFÍ*

> You ask me of my impressions of the peace settlement.
> How can I hide the depth of the melancholia with
> which I sign the Treaty of Lausanne[?]
> —Venizélos, July 24, 1923

The Allied victory in 1918 was also a political victory for Venizélos. Over
the course of the war he had become popular with the great powers, and
as a champion of Wilson's vision for the League of Nations he made powerful friends and was given a large say at the bargaining table during peace
settlement negotiations. Even to the royalists, who continued to despise
Venizélos, it was obvious that Greece's best interests would be served
by keeping him at the political forefront. Largely because of Venizélos's
popularity, the postwar settlement gave Greece territory disproportionate
to the fighting the Greeks had actually done for the Allied cause. In a huge
vindication of Venizélos's unflagging allegiance to the Great Idea, the
1920 Treaty of Sèvres gave Greece all of Thrace and most critically—and
ultimately most calamitously—Smyrna on Asia Minor's coast, along with
its surrounding territory.[79] Greece would now be a country "of two conti-
nents and five seas," in the rhetoric of the day. The territory was to be
handed over sequentially to Greece in a series of plebiscites.[80] While all
of this looked good on paper, in reality it was little better than written on
ice. Of the great powers, only the British gave wholehearted support to

Greek territorial expansion into Asia Minor. Nevertheless, Venizélos urged Greece, exhausted from an unremitting series of military campaigns—the Greco-Turkish War, the Balkan Wars, and World War I—to press on to actualize the promises of the Paris Conference, landing troops in Smyrna in May 1919.

What followed was a period of unprecedently tumultuous domestic politics, even by Greek standards.[81] In the meantime, just across the Aegean the Turkish national movement, led by Mustafa Kemal Attaturk, a Salonikan Turkish nationalist, came into existence. Attaturk rejected the Treaty of Sèvres and defied Greece to attempt a takeover of Smyrna. Britain, so recently the strongest supporter of Greek postwar territorial expansion, now began backpedaling in light of French support for the Turks, and the realization that fulfillment of the terms of Sèvres would lead to major conflict between the new Greek and Turkish governments.

On March 23, 1921, the Greek military began a major offensive in Anatolia. For the first several months it looked successful, with the Greeks moving farther and farther into the Anatolian heartland. But the deeper they got, the more vulnerable they became. In August 1922 Kemal launched a devastating counteroffensive, utterly destroying any semblance of order or planning on the Greek side. The Greek army fled in chaos, heading for Smyrna on the coast, where the Greek population outnumbered the Turkish by a ratio of two to one.[82] Before the mass arrival of the refugees, there were about 150,000 Greeks living in the city, almost half its total population.[83] The runaway Greek forces filled the city. "In a never-ending stream [they] poured through the town toward the point on the coast at which the Greek fleet had withdrawn. Silently as ghosts they went, looking neither to the right nor the left. From time to time, some soldier, his strength entirely spent, collapsed on the side-walk or by a door."[84] Along with them came thousands of Greek Orthodox civilians fleeing Turkish reprisals.

The status of Asia Minor Greeks had become increasingly tenuous since the Young Turk revolution. The Greek military campaign in Turkey dangerously intensified their vulnerability. On June 14, 1915, the Turkish government announced its decision to begin the forced conversion of Christians through mixed marriages; in actuality the result was mass deportations of the Christian populations from Marmara.[85] The *Morning Post*'s Turkish correspondent reported that the Kemalist government had as its aim nothing short of the "extermination of [the Greek] race."[86] M. Henry Morgenthau, the U.S. ambassador to Constantinople, wrote that the Turks had "adopted almost identically the same procedure against the Greeks as that which they had adopted against the Armenians"—using them in forced labor battalions, and torturing women and children.[87] While less well documented, the Greek troops also committed atrocities against the local Turkish population. These included the rape of children and old women, the desecration of mosques, and the torture and murder of civilians.[88] The Greek "liberating" forces, in their initial position of relative strength, had committed atrocities against the Turks that the Turks now sought to avenge.

The situation in the city became more and more dangerous. When the Turkish forces arrived on September 9, 1922, a slaughter ensued. First they attacked the Armenians, and then the Greeks. The archbishop was set on by a mob, and civilians cowered in their homes waiting to be murdered. Eight years earlier, Venizélos had warned that the Greeks of Asia Minor would suffer the same fate as Turkey's Armenians.[89] The Greeks of Asia Minor had long feared this as well.[90] Now, in Smyrna, the prediction came true. On September 13, a fire broke out in the Armenian district. It quickly spread, and tens of thousands of people, mostly Greek Orthodox, rushed to the waterfront hoping to be evacuated. They were not. Almost all of them perished.[91] Any chance of Greece's ever gaining Asia Minor was gone.[92] After a last Greek offensive in Thrace also ended in defeat, Venizélos capitulated.[93] Turkey was the definitive possessor of Asia Minor, Constantinople, and even eastern Thrace. The Great Idea had run its course.

The *Megáli Idéa*'s territorial claims had long rested on a significant Greek Orthodox presence outside of Greece's borders. The settlement took care of this as well. Under the terms of the Treaty of Lausanne, all Muslims living in Greece (except those in Thrace) were to be compulsorily relocated to Turkey; all Orthodox Christians in Turkey (except those in Istanbul) were to go to Greece.[94] About four hundred thousand Muslims left Greece, and almost one and a half million Orthodox Christians were sent to Greece. In this regard, one aspect of the *Megáli Idéa* was upheld: the equation of Greekness with Orthodox Christianity. Many of the refugees who came to Greece spoke no Greek, and were fully Turkish in their cultural practices; religion was the sole basis on which they were designated as Greek.

Some postwar observers argued against such an arrangement, positing that "the Turks and the Greeks will united make a happier country than either race could by itself," but the reality on the ground already had determined that the two were to be separated for good.[95] Refugees in the hundreds of thousands had been pouring into Greece since the last months of the Asia Minor campaign. With the additional formal exchanges entailed by the Treaty of Lausanne, Greece's population increased by one-fifth within the space of less than a year.[96] The horrors of the situation were most acute for the refugees, especially the vast majority who had fled in the wake of the war, rather than under the relatively more ordered circumstances delimited by Lausanne.[97] The Asia Minor Greeks, many of whom had witnessed firsthand the Turkish atrocities against Armenians, had suffered immensely during the period; in some regions, there were reports of hundreds of Greek women and girls drowning themselves so as to avoid torture and execution at the hands of the Turks.[98] But the stresses were also massive for the areas that received the refugees, especially Athens and Salonika. Both cities, but particularly Salonika, were already suffering from economic collapse, political unrest, and the fallout from a series of military campaigns that had gone on more or less unabated for over a decade. With the arrival of hundreds of thousands of refugees in cities where resources were already scarce, those who were the most vulnerable suffered the most.

Overwhelmingly, the task of absorbing the newcomers fell on Macedonia. As the Refugees Settlement Commission explained, "Macedonia is the district in which the great bulk of the refugees will be settled, because it is potentially one of the most fertile districts in Greece. . . . [It also] has a large number of Turkish inhabitants who, under the terms of the Convention for the Exchange of Populations (Lausanne Treaty), are being removed to Turkey."[99] In theory, the departure of the Muslims of the region would free up space for the incoming Christians. But the recent devastation of Salonika proved useful as well.

The influx of refugees coincided with the city's reconstruction, the beginnings of a tentative Jewish outward migration, internal battles over property, and the desperate attempts of thousands of residents to regain their economic footing. Fund-raising campaigns were launched regularly each year at the Jewish holidays by the Executive Committee of the Zionist Organization.[100] Economically, many were stretched to the limit; the competing demands of various charitable causes could not be met. The arrival of the refugees also coincided with an increasingly overt campaign on the part of the Greek government to undo Jewish hegemony. The 1917 fire had presented Greek authorities with the occasion to rework the physical and cultural geography of the city. Now, hundreds of thousands of refugees would be the instrument of demographic change. Refugees of Smyrna's fire became competitors with those uprooted by Salonika's. The city would scarcely be able to accommodate so much resentment, misery, and anger.

The Asia Minor refugees suffered ugly prejudice from the Greeks who "welcomed" them on arrival. Nazos Kyriakopoulos sarcastically recounts his arrival in Patras after a weeks-long journey: "What wonderful people they were! Callous, very callous, the people of Patras. Hungry, tired, and weary from the journey, we heard them welcome us: 'What do you want in our country, sons of Turks? Go and see your friend Venizélos.'"[101]

The situation in Salonika was particularly tense. Jews there felt that the Greek newcomers had been intentionally and explicitly pitted against them, which in many cases seems to have been the case. While accounts from around Greece indicate that the treatment Kyriakopoulos received at the hands of Patras Greeks was typical of patterns throughout the country, the Jews of Salonika felt that only they were singled out by Venizélos as unwelcoming of the refugees. As Yitzhak Immanuel recalls,

> The new Greeks in the city received a lot of special privileges. They were allowed, for example, to sell their goods on the doorsteps of the Jews' stores. Since they were exempted from taxes and fees [and since] they didn't pay rent or have other expenses, they gave tough competition to the Jewish merchants. Even though the Jews didn't protest, even donating money for their relief, Venizélos complained that "the Jews did not welcome the refugees with open arms."[102]

HELLENIZATION AND ANTI-SEMITISM

It is difficult to determine whether the relationship between Jews and the newcomers was particularly strained, or if the tensions were not much different from those experienced between "old" Greeks and the new arrivals. Given the relatively calm history of Christian-Jewish relations in Salonika in the nineteenth century, particularly in conjunction with the high rate of cases of blood libel and other anti-Semitic incidents in Asia Minor in the same period, it seems safe to conclude that Asia Minor Greeks were more hostile toward Jews than were their Salonikan Christian counterparts—who after all, unlike Asia Minor Greeks, had cohabited with large numbers of Jews for centuries. Salonika's Jews certainly felt that anti-Semitism was particularly characteristic of Asia Minor Greeks and felt unjustly singled out as unaccommodating of the newcomers.

The influx of refugees, while creating huge economic and social difficulties for the Greek government, worked hand in hand with its policies of Hellenization; even as the refugees arrived, Jewish Salonikans were already struggling with a slew of legislation targeted at reducing their influence. In some cases, Hellenization was clearly linked to anti-Semitism; in others it was not. On the one hand, Hellenization can be seen as a natural outcome of Greek national expansion and decades of Greek anxiety over the fact that "naturally" Greek territories lay outside the state. With the recent inclusion of those territories, proponents of the Greek national project were eager to justify its long-standing claims. At the same time, much of the Hellenization of the period was as anti-Jewish as it was pro-Greek. The newspaper *Makedonía* published vituperatively anti-Semitic articles and at times had to be checked by the Greek government. The Jewish community stepped up its political activities, with the

Jewish Congress of Greece meeting for the first time on March 11, 1919.[103] The extraordinarily diverse Salonikan press, which included publications in Ladino, Hebrew, Greek, Turkish, English, Italian, French, and Russian, proliferated in this era of fraught politics, with papers being founded to espouse one or the other cause.[104]

In 1922–23, the Salonikan municipality forced out all Jews working in the port. This was the final blow after earlier attempts to limit their influence at the docks. The fired workers were given some compensation, but had no chance of reemployment. Asia Minor Greeks were hired in their place. All Jewish livery drivers were also prohibited from continuing their work. Jews engaged in fishing, too, were slowly pushed out of their business. The Greek military commander of the city, Pangalos, was particularly cruel to the Jews, and used specially enacted "emergency laws" to force them from the economic sphere.

In 1923, two Jewish scrap metal merchants were accused of tampering with the city's telephone lines and sentenced to death. Despite the interventions of community leaders, the two were executed in March 1923. Jewish taxpayers came under especially close scrutiny and had to pay disproportionately high taxes. Finally, Ladino and Hebrew were banned from all public signs. The culmination of these developments was a Jewish boycott of the December 1923 elections, which protested Venizélos's redistricting of the city, clearly designed to limit the influence that Jewish voters—who were overwhelmingly royalist—would have on the elections.[105] The new redistricting remained in place until revoked by Ioannis Metaxas in 1934.[106] Venizélos, if largely for his own political reasons, was the champion of these anti-Semitic measures. Although he issued a declaration that Jews were to have the same freedoms and civic rights as the rest of the Greek citizenry, he was widely perceived by Salonika's community as an enemy.

The question of the Asia Minor refugees, and the particular tensions of their relationship with the Jews, is interwoven with this broader picture of increasing Jewish vulnerability, marginalization, and resentment. Developments of the day, anti-Semitic or not, were quite understandably interpreted by the Salonikan Jewish community as having specific relevance and reference to them. On September 29, 1923, Great Britain, France, and Italy established a Refugees Settlement Commission to assist the Hellenic government. It was also to serve as a clearinghouse for property and restitution claims. Of the commission's four members, two were appointed by the Greek government and one by the League of Nations. The fourth—the commission's chair—was to be a U.S. citizen.[107] That the appointed chair of the committee, Henry Morgenthau, was Jewish was significant to many Salonikan Jews, who hoped that this fact would lessen accusations that the Jews as a whole were unkind to the Asia Minor refugees. "The Jew Morgenthau, Treasury Minister of the U.S., arranged a loan of ten million pounds sterling for relief for the refugees, but even that did not appease the anti-Semites."[108] In reality, though, Morgenthau's position derived from his experiences in Istanbul, not from the fact that he was Jewish.

While one article of Lausanne had addressed the legal status of some Jews (those without Ottoman nationality who settled in Palestine), the treaty's overwhelming preoccupation in terms of minorities was with Orthodox Christians and Muslims.[109] The general wording assured the protection of both Greek and Turkish minorities, but again, this was designed primarily with Christians and Muslims in mind: the Christians who remained by special arrangement in Constantinople, and the Muslims of Thrace. Jews in Greece were regarded as Greek citizens, not as minorities in crisis. Lausanne was suffused with the sense on the part of the great powers that enmity between Greece and Turkey had been dangerously allowed to go too long unchecked. Its resolution, and not the plight of minorities, was the central goal.

Lausanne provided a blueprint for future diplomacy in the Mediterranean. Many of Salonika's Jews, who had been increasingly marginalized with the arrival of hundreds of thousands of Orthodox refugees from Asia Minor, would next hear of the exchanges in quite a different context. In the 1930s, when Britain proposed the partition of Palestine, a component of which was to be a "transfer" of populations, Lausanne was held up as a shining example of success.[110] While many Zionists agreed with this position—and while Arabs condemned it—Salonikan Jews who had immigrated to Palestine in the wake of Lausanne were uniquely positioned to understand its implications. Lausanne was a product of a World War I environment that also produced the Balfour Declaration, another redrawing of the demographic and geographic map that had later significance for some Salonikans.[111]

In 1912, the population of Salonika had been roughly 30 percent Greek Orthodox, 25 percent Muslim, and 40 percent Jewish. By 1926, it was 80 percent Greek Orthodox and 15 to 20 percent Jewish. In Macedonia as a whole, 513,000 Greek Orthodox Christians had made up roughly 43 percent of the population; by 1926 they numbered 1,341,000—fully 89 percent.[112]

In effect, Greece had exchanged a territorially broad but diffuse Hellenism for a small but concentrated one. In 1928 one observer commented, "What [Greece] lost in extent in Asia, she gained in intensity in Europe."[113] While the total number of Sephardic Jews in Greece remained constant through the period—roughly seventy thousand both in 1913 and in 1928—in percentage terms they dropped off precipitously.[114]

The Greek government fully understood that the arrival of hundreds of thousands of new Greek Orthodox citizens provided an opportunity to resolve the long-standing problem of Macedonia's ethnic homogeneity. In 1922 Venizélos wrote,

> The very future of Greece is dependent on the success or failure of the solution of the refugee question. A failure would cause many calamities, while a success would allow Greece to recover in a span of a few years from the burdens bequeathed by the Asia Minor Catastrophe. After the collapse of Greater Greece, we can consolidate the borders of Great Greece only when Macedonia and Western Thrace have become not only politically but also ethnically Greek lands.[115]

Just one example of how much things changed as a result of the shift in demographics is that in 1919, when Venizelist legislation made Sunday the official day of cessation of business in Salonika, the Jewish community protested so vigorously that the law was abolished after one week. Five years later, right after the population exchanges, similar legislation was again enacted—despite Venizélos's promise that it would not be. This time it was successful, despite an appeal to the League of Nations.[116] With the arrival of more than one hundred thousand Asia Minor Greeks in Salonika alone, the city's Jews no longer made up nearly so significant a proportion of the population, and the Judeo-phobic attitudes of the incoming refugees worked to further marginalize them.

Forced, rapid social change in Salonika was closely linked to the fact that Jews inhabited the city in such number. In other areas, where there were few Jews, it would have been nonsensical to pass legislation dealing with specifically Jewish matters. Legislation dealing with Salonika often was designed to target its Jewish population. But the aggressive measures taken in the city by the Greek government were not simply an anti-Semitic reaction. They should also be read against the broader backdrop of the city's early twentieth-century history. Long-standing yearning for Macedonia fused with the turmoil of the First World War and the virtual civil war; the great fire paved the way for overt social intervention. Other factors, among them the rise of international Jewish movements and organizations, provided an explanation (or more cynically, a pretext) for the Greek government's reluctance to show any preferential treatment toward Greek Jews as such. Finally, the Salonikan Jewish community's desire to retain Ottoman legal structures and imperial privileges was in conflict with the uniformity of Greek legislation. A contemporary British observer argued that "the fact that there is one law for all . . . constitutes the Salonika Jew's [sic] grievance."[117] This is an oversimplification. Salonika's Jews had a number of grievances, not all of which rested on the desire for special or separate treatment. But the extent to which the mood of the Salonika community in the early years of its incorporation into Greece stemmed from the difficult transition from an imperial to a national framework should not be overlooked. The shift from Ottoman to Greek rule was an excruciatingly complicated and problematic one for Salonika's Jews. That this was so, however, derived as much from the disappearance of empire as from the imposition of the Greek nation-state in its place.

PART THREE

Normalization to Destruction

Chapter 6

INTERWAR GREECE: JEWS
UNDER VENIZÉLOS AND METAXAS

IN A RECENTLY published memoir, Moisis Vourlas, a Greek-speaking Jew born in Egypt in 1918, writes of his family's return in the early twentieth century to what he calls "mother Greece" (*mitéra Elláda*). Vourlas tells the story of a homework assignment he was given as a youth, in the 1920s after his family's return "home":

> As a homework exercise, the teacher had us write an essay on the topic of "homeland." . . . I sat down in the afternoon, and from two until midnight, with the help of my two older sisters, who remembered better than I did our life in Egypt, I wrote a long account. . . . In the beginning I described Cairo . . . and the life there . . . the customs of the Greeks and the Jews, but then I recalled everyone's nostalgia [*i nostalgía mas*] for mother Greece and I wrote of my great joy that I had come home again.[1]

The Salonikan schoolteacher who came up with the assignment doubtless had it in mind to target the students in the class—of whom in the late 1920s there would likely have been many—who had recently moved to the city from Asia Minor. What did *patrída* mean to them? But here was Vourlas, an Alexandrian Greek Jew, sitting beside them. When he turned his mind to the question, he first thought of Egypt, but on reflection embraced Greece, instead, as his real homeland. It's likely that this is the answer the schoolteacher hoped to prompt from all of the students. Vourlas's phrasing is nevertheless revealing: mother Greece was home to him, a place not where no distinction was made between Greeks and Jews, but a place that could rightfully be claimed by both as home. This formulation was in keeping with the new Greek Jewish culture that emerged particularly in Salonika and Athens in the 1920s and 1930s—a nascent but inchoate Greek Jewish culture that would not have the chance to develop to the point of fully understanding the relationship between Jewishness and Greece.

Only after the annexation of Salonika did the Greek state have to come to grips, in legal terms, with the status of the nation's minority citizens. The various constitutions of the nineteenth and early twentieth centuries, of which there were six (1822, 1823, 1827, 1844, 1864, and 1911), had made scant mention of minorities, and as the century progressed were increasingly inclined toward the hegemony of the state religion, Greek Orthodoxy. While the earliest constitution had affirmed the Eastern Orthodox Church as the "dominant" religion but emphasized the state's tolerance of all other religions, from 1844 on all the constitutions reserved certain rights—proselytizing, for example—to Orthodoxy alone and were fuzzy as to the prospects non-Christians in the kingdom had for attaining full citizenship rights equal to those of Greek Orthodox Christians.

With the annexation of Salonika, this legal imprecision was clearly no longer viable. While up until 1912 Jews had lived in scattered, largely provincial pockets around the country, now the nation's second-largest city was home to tens of thousands. The Salonikan community was aggressive in lobbying Greek Parliament for clarification of the legal status of Jews in Greece, with the result that from 1920 on, to be Jewish in Greece was not simply, as Bernard Pierron has put it, "to belong to a minority 'ethno-confessional' group, tolerated so long as it remained quiet," but rather to have the right to expect one's difference to be respected and protected by law.[2] This change in Greek legal perspective proved to have a prescriptive, rather than merely descriptive, dimension: with the recognition of Jews as rights-bearing citizens, Jewish identity gradually expanded to encompass more than solely the "religious" (Pierron's ethno-confessional) sphere. That is, to be a Greek Jew was as much a matter of Greek citizenship as of Jewish confession. The emergence of Greek Jews, as a category recognizable to both the Jews whom it described and the Orthodox Christians beside whom they lived, was not, then, a matter simply of assimilation. To become Greek did not necessarily mean an abandonment of Judaism. Rather, it entailed an expansion of Jewish identity to encompass not only a religious but also a civic status.

Law 2456/1920, adopted July 27, 1920, recognized Jews as a protected minority citizenry and for the first time allowed, from both a Greek Orthodox Christian *and* Jewish perspective, for the conceptualization of a category called Greek Jew. Indeed, from 1920 the Salonika community began referring to itself in its formal communications not simply as the Israelite Community of Salonika but also as the Greek Jewish Community. (As one obvious result, all Jews in Greece were increasingly drawn together; with the creation of a juridical category to which all Greek Jews belonged, the particularity of the Salonikan community vis-à-vis other Greek Jewish groups was reduced.) This legal development, more than the simple fact of annexation to Greece, provided a catalyst for the emergence of a distinctly *Greek* Jewish identity. Thus, for example, in the 1930s an important amendment was made to the law of 1920, making Greek citizenship a requirement for official participation in the life of the community. The grand rabbi and other rabbinic appointments had to hold Greek nationality to be eligible, as did all participants in the Jewish Community Assembly.[3] Over the course of the 1930s a series of similar refinements were made, along with the formalization of the status of various Greek Jewish communities (ultimately, thirty-one of them) around the country. Most legislation regarding Jews in the period dealt with internal matters, such as the fine-tuning of community structures, the official titles of religious functionaries, and the requirements for holding community positions. Some—particularly legislation that was, or seemed, designed to limit the influence of Jews in Greek economic and political matters— spurred conflict and negotiations between the Jewish community as a whole and the Greek state. But little of it indicates a sense on the part of Greek Jews that they needed special "protections" from the Greek state or Greek Christians.[4] In short, most legislation of the interwar period that specifically addressed Jews had as its outcome, if not purpose, the reconfiguring and redefining of the community within a national context—establishing what, juridically and politically, it meant to be a Greek Jew—and was not premised on antagonistic relations between the Orthodox Christian and Jewish communities.

With the cultural assimilation of thousands of Jews, a true Greek Jewish culture took root. Its best-known symbol is perhaps the famous Rembetika singer Rosa Eskanazi, but an array of other Greek Jews, too, gradually made inroads into the mainstream of Greek culture. Indeed, the Rembetika genre as a whole—first championed by the Asia Minor refugees—which blended Oriental and Western styles, and was devoted to telling the story of society's underdogs, was closely linked to Greek Jewish life. In 1924, the German recording company Odeon appointed the Salonikan Jewish firm Abravanel and Benveniste as "their exclusive agents for the purposes of artist and repertoire selection."[5] Sephardic musical traditions—particularly the music of women singers known as tañaderas—were to a degree integrated into this now most distinctively Greek genre, Rembetika.[6]

By the time of the episode described by Vourlas, a first generation of Greek-speaking Salonikan Jews had begun to form. After the Greek conquest of Salonika, Greek language instruction was mandated in all schools that enjoyed public funding. In many instances, the Greek government provided funding to Jewish schools specifically for the teaching of the Greek language.[7] Just as in immigrant contexts there is a rapid shift in language from the first generation to the next, in the new Greek Salonika change came rapidly after the transition from Ottoman rule, which rendered the city's citizens de facto, if unwilling, immigrants to Greece. By the late 1920s, the younger generation of Salonikans was largely bilingual; by the 1930s, many had Greek as their first language. Greek Jews began writing in Greek, with a small body of Greek Jewish literary works emerging in the period. The choice of language, of course, was political, and Greek was not universally embraced—with elites favoring French over Ladino and Greek, and Zionists emphasizing the need for Hebrew instruction. Among the younger Sephardic middle classes, however, Greek had established itself as the primary language by the 1930s. Some of the most poignant documents of the Greek Jewish experience of the Holocaust were written in Greek: hastily penned farewell notes, songs about life under the Nazis set to the tune of Greek hits of the 1930s, and memoirs of the few survivors. Given time, Greek Jewish culture might well have asserted itself as a readily recognizable feature of the Greek cultural landscape as a whole.

MACEDONIA AND THE CAMPBELL RIOTS OF 1931

This is not to say that Greece was bereft of anti-Semitism. Far from it. Much of the nationalist press struck up the repeated theme, most stridently the Salonika newspaper *Makedonía*, but also the Athens daily *Akropolis*. A scattering of ultranationalist organizations, most prominently the Ethnikí Énosis Elládos (EEE, or National Union of Greece), flourished in the period; to this day such groups in Greece are strongly anti-Semitic. Founded in 1927, the EEE's membership numbered in the thousands, and was made up largely of Asia Minor refugees. The refugees, so recently themselves the target of xenophobia, now proved to be a highly conservative political bloc. The fact that the one thing they clearly had in common with most Greece-born Greeks was their Christianity placed Orthodoxy at the fore of their understanding of what it mean to be Greek. They, newcomers, emphasized their "Greekness" through the language of Orthodoxy.[8] Salonika's Jews, while resident for generations, appeared to them as foreign interlopers.

Effectively, the EEE's media mouthpiece was *Makedonía*, which had an overtly and rabidly anti-Jewish editorial policy. The paper editorialized daily on the "anti-Hellenism" of the Jews and complained of the Zionist opposition to the education of Jewish children in Greek schools, among other issues. Increasingly, Greece's anti-Semitic press singled out the Jews as traitors who were not really Greek, suggesting that the territories added in 1912–13 were threatened by them. Typical is an October 1929 editorial in *Makedonía*: "Either [the Jews] will acquire a Greek consciousness, identifying their interests and their expectations with ours, or they will seek a home elsewhere, because Thessaly is not in a position to nurse at its bosom people who are Greeks only in name."[9] The Ladino Communist daily, *Avanti*, retorted that the EEE should be renamed Éllines Exontósate

Evraíous, or "Greeks, Eliminate Jews."[10] Just as the Asia Minor Greeks had feared extermination at the hands of the Turks, Greek Jews now fretted over their future in Greece.

The Protocols of the Elders of Zion, printed in a Greek edition by Aristides Andronikos in 1928, enjoyed wide circulation, and Andronikos published articles on the "Jewish threat" with an obsessive frequency. Even as Venizélos gave public assurances to the Jewish community that such views reflected only an extreme fringe and had no connection with the official stance of the government, he himself was well-known as still harboring anger over what he perceived as Jewish disloyalty to Greece during the Balkan Wars. Indeed, Venizélos and the Liberal Party were no strangers to anti-Semitism. His exceeding intolerance of the Greek Communist Party, despite the fact that it regularly saw election returns only in the low single digits, was not unconnected to its proportionally large Jewish membership.[11] When in 1929 he passed regressive legislation that severely hampered the rights of political dissenters to express their views publicly, Jews saw themselves as a target.[12] And when in 1929 the Jewish organization Mizraḥi endeavored to have the 1924 law on Sabbath observance—which had declared Sunday the official day of rest—revoked, its request was ignored. In Genoa for the 1930 League of Nations convention, Venizélos assiduously avoided Jewish advocates who tried to lobby him in the matter on behalf of Salonika's Jews, and enraged them by saying of the legislation, "My impression is that there is no problem."[13] In interviews with the press, he was outspoken about the fact that his support of Zionism was based on the hope that it would encourage Jews to leave Greece, and increasingly toward the end of his life he spoke freely about his ongoing anger at their "betrayal" in 1912.[14]

The reality was that interwar Greece was characterized by what Pierron has insightfully identified as a peculiar bifurcation in Greek attitudes toward Jews in the period.[15] At once eager to embrace the language of equality and liberal democracy, and distinguished by a vocal ultranationalist fringe, Greece to this day is characterized by a formally liberal democratic state that lives close at hand to vituperative anti-Semitism, usually sponsored by marginal nationalist groups, but also by church leaders and even political leaders themselves. It is not easy to characterize interwar Greece in blanket terms as anti-Semitic or tolerant, welcoming or dismissive of its Jewish citizens.

That the city of Salonika somehow managed to stumble through the 1920s with no major intercommunal violent outbursts is, seen in one light, quite remarkable; with upward of a hundred thousand new Asia Minor Greeks being told that in leaving Turkey, their ancestral home, they had "come home," even as sixty thousand Jews who'd lived there for generations were increasingly marginalized, the potential for calamity was huge. The global economic crash of 1929 had a tremendous impact on Greece (by 1933 the nation was defaulting on interest payments to foreign lenders), and by the end of the year close to two-thirds of the city's population was reliant on state welfare.[16] The older generation of Salonika's Jews could scarcely function in the national language. Litigation over property claims related to the fire of 1917 was ongoing. The key factor, however, that would ignite the situation, bringing simmering tension to the level of outright violence, was an old one: Greek claims to territory beyond the borders of the state.

The definitive status of Macedonia was not resolved by the Balkan Wars. Bulgaria, Greece's ally in the first Balkan War and its foe in the second, refused to accept the new Balkan status quo and relinquish its claims to Macedonia. In the thirty years between the Balkan Wars and the Second World War, Bulgaria provoked a series of conflicts on the northern Greek frontier, coming to be regarded by the early 1940s as the "Germany of the Balkans," destabilizing the region with its constant aggressions. As one journalist observer put it, "For the Balkan peoples there is a problem of Bulgaria, as there is a problem of Germany for the whole world. . . . [Bulgaria] has in the space of twenty-eight years unleashed three wars on her neighbors causing them untold suffering."[17] After the population exchanges (which had included an exchange between Bulgaria and Greece in 1919), Bulgaria's recalcitrance provided a foil against which Venizélos's willingness to reconcile with Turkey appeared all the more noble, and which largely erased the West's recent annoyance with Greece over Greece's own earlier aggressions.[18] "The scowling Bulgaria sat back with dark thoughts of revenge instead of learning to live in peace with her neighbors," noted the same journalist.[19]

On the eve of the Balkan Wars, Salonika's Jews had not been enthusiastic about Greek territorial control of the region; their efforts to gain it autonomous status after the annexation were well-known. Now they were suspected of still harboring similar sentiments regarding the final disposition of Macedonia. A complicated set of factors contributed to this. Among them was the increasing popularity, albeit modest, among Greek Jews of the Greek Communist Party (Kommunistikó Kómma tis Elládos, or KKE). Founded in 1918, the main obstacle to significant growth of the party among the Greek mainstream was not ideological but rather the result of the Comintern's unwavering support for an independent Macedonian state[20]—a position at diametric odds with the Greek nationalist view that as Greek propaganda has it to this day, "Macedonia always has been, is, and shall be Greek."

Even before the official founding of the party, Communism had attracted the attention of Salonikan Jews, particularly among the younger, alliance-educated generation. In 1914, the Ladino-language series *Bibli-*

oteca Communista published as its second work a Ladino-translation of Marx, *Il Manifesto Communista di K. Marx und F. Engels*, translated by Yitzhak David Florentin. Many previously socialist Jewish groups migrated leftward on the political spectrum. In 1924, for example, the editorship of the previously socialist newspaper *Avanti* was taken over by Jacques Ventura, one of the leaders of Salonika's Communist Party. At a time when all good Greek nationalists were expected to support a Greek Macedonia, international Communism's position on Macedonia rendered its Greek supporters suspect. This was doubly the case for Jews, whose recent alliance with the Ottomans over the Greeks in the Balkan Wars was still held against them, especially by the Greek press, to such a degree that one Salonikan paper, *Makedonía*, had an editorial stance that was virtually consumed by the idea of the antinationalist perfidy of Greek Jews. Founded in 1909, from the start the paper had as its primary editorial orientation a nationalism that was all but coterminous with anti-Semitism. After 1928, its depictions of Jews were taken more or less directly from the *Protocols*: Jews planned to take over all Greek institutions; they had infiltrated the economy; they were secretly running the state. In summer 1930 the Greek Jewish community of Salonika appealed to Stylianos Gonatas, the governor of Salonika, requesting that Jews of non-Greek origin be allowed to continue official participation in the life of the city's Jewish community. When the request was upheld, *Makedonía* published a series of vituperatively bitter articles on the anti-Hellenism of Salonika's Jews. At the core of the issue, of course, was anti-Semitism. But another feature of the situation, as in so many other intercommunal matters over the course of the nineteenth and early twentieth centuries, was a fundamental lack of clarity on all sides about what national citizenship meant. Embedded in the Jewish request that non-Greeks be formally recognized as members of the Greek Jewish community was a paradox—one that derived from the difficulty with which not just Greeks but Jews too relinquished community-based imperial categories.

Later that year, a Salonikan delegate to a Zionist convention in Bulgaria was condemned in the Greek press for not protesting a Bulgarian speech arguing in favor of a Bulgarian Macedonia.[21] But *Makedonía* had yet to produce evidence of Jewish betrayal so great that it would inspire large numbers of Greek Christians to turn against them. While filled with innuendo about Jewish perfidy, betrayal, and anti-Hellenism, the paper had no smoking gun. But in 1931, it seemed (wrongly, as it turned out) that it had. In June of that year Yitzhak Koen, a member of the Zionist Salonikan Jewish sporting organization named Maccabi, traveled to Macedonia for a regional meeting of the organization commemorating its twenty-fifth anniversary. At about the same time, a Bulgarian nationalist conference

was also meeting in Sofia, at which a resolution for Macedonian autonomy was adopted.[22] When *Makedonía*'s journalists learned of Koen's trip,
they accused him in print of having gone to attend the conference.[23] (By
some accounts he did attend it, but left before the resolution was passed;
by others he had not been there at all.)[24] In past years, *Makedonía* had
published countless articles in which national sentiment had been a fig
leaf for anti-Semitism; now all pretense disappeared. Over the next four
days, from June 20–24, *Makedonía*, along with various ultranationalist
organizations, called for the disbanding of the Maccabi association and
accused the Jews in blanket terms of treason against the state.[25] Despite
calls for calm from other quarters, anti-Semitic violence inevitably broke
out on the evening of June 24, when a band of close to one hundred
ultranationalist youth descended on Maccabi's Karaiskaki headquarters,
sacked it, and wounded a number of the members who happened to be
there. In a tableau reminiscent of Pacifico's plight almost a century earlier,
the authorities took close to an hour to arrive on the scene, by which time
the perpetrators had largely dispersed.[26]

By the end of the week, the situation had only deteriorated; with
Makedonía and even the mainstream Athenian press heaping the blame
on the Jews, Salonikan Jews feared for their lives, staying inside with their
businesses and homes shuttered. Signs were posted reading "Jews get out
of Salonika." Riots led by Greek nationalists demonstrated for the expulsion of the city's Jews, and "represented [them] as being mostly foreign
subjects making money in Greece and cooperating with communists and
comitadjis [Bulgarian fighters]."[27] As one Jewish eyewitness later recalled,
"Anti-Semitic gangs roamed the city and marked the houses systematically with the letters 'K' or 'H,' announcing that they belonged to Jews.
In addition, anonymous threat letters were sent to many Jews, telling them
to leave the city."[28] On June 29, *Makedonía*'s oversize headline urged its
readers to "Finish [the Jews] Off!" That night, a crowd numbering in the
thousands set fire to the heavily Jewish "Campbell" district, a poor area
where Jewish refugees of the 1917 fire had been resettled in dwellings that
were little more than huts.[29] Quickly, the violence spread to other districts
as well, with synagogues laid siege, houses burned and looted, and Jewish
property destroyed across the city. Venizélos now moved swiftly: he dele-
gated government officials to meet with Jewish leader, and reiterated to
Parliament that "the attacks on the Jews were lamentable and without
justification."[30] Martial law was declared, and an army battalion was sent
to protect Salonika's Jewish neighborhoods.

The episode provided, to a small extent, the occasion for some soul-
searching. Representatives from various "professional and commercial
organizations," Jewish and Greek, held a conference to express and attempt to bridge their differences. The Venizelist government gave public

declarations that Greek Jews were to be treated no differently than other Greeks, and moved within one week to allocate a half million drachma for the reconstruction of the destroyed homes.[31] But this flurry of bridge building was short-lived. In the trial that followed, which took place in Veroia in 1932, the accused were all acquitted; while the court ruled that they had in fact committed arson, it argued that the perpetrators had been motivated by patriotism and had honestly felt that the Jews posed a threat to the homeland, so should not be punished for their actions. Patriotism, in the court's view, was all that really had been at play, and it was not a crime.

The fear engendered by the Campbell riots, and the disgust engendered by the subsequent trial, triggered a modest wave of emigration. The outcome of the trial, particularly, was devastating. While the riots could be ascribed to a group—albeit sizable—of fanatics, not representative of Greek opinion as a whole, the opinion of the court seemed a formal, state-sponsored valorization of anti-Semitism. Although the protesters had directed some of their animosity at the French (French representatives had also been at the meeting in Sofia, and some extremists had briefly planned to set the Mission Laïque Française on fire), most of their hatred had been reserved explicitly for "the Jews."[32] Now the failure to acquit them seemed a formal endorsement of the view that Jews were not really Greek. This feeling was only corroborated over the next years as a series of municipal legislative measures—much of it designed to give the city control of Salonika's vast Jewish cemetery—was enacted. By 1936, when the fascist dictatorship led by General Ioannis Metaxas came to power, Salonika's Jews were a community under siege and in transition.

In the late nineteenth century, the Greco-Turkish War had been a watershed moment for Greek Jews—an episode that rapidly undermined the hard-earned, wary sense of security that had gradually emerged among many of Greece's Jewish communities in the late 1800s. The Campbell riots had much the same impact on Salonika's gradually emerging Greek Jewish consciousness. In its midst, a highly regressive, latent anti-Semitism emerged, and rigid communitarian lines dividing "Greeks" and "Jews" were redrawn. As one contemporary observer described it, the "original Nationalist-Maccabi controversy . . . bec[a]me a Greco-Jewish issue," which many Orthodox Greeks read as "a holy war for the defence [sic] of their children."[33] The Maccabi riots became the occasion, yet again, for an expression of the idea that Jews weren't really Greek, no matter their citizenship. They were foreigners, Communists, or proxy Bulgarians; they were unpatriotic traitors and enemies of Hellenism. Since 1821, Jews in Greece, or in lands claimed by Greece, had heard a similar refrain. It is unlikely that the setback would have been any more permanent than would be earlier ones. But by 1931

Greece's Jews—particularly in Salonika—were under extraordinary pressure from both without and within.

Under the cumulative weight of a dizzying variety of events, the Jewish community was imploding ideologically as well as declining physically. Debates over Zionism had already been heated in the late 1920s. While *La Renacensia Giudia*, the Zionist paper, argued against the use of the Greek language, *El Pueblo*, another Jewish paper, upbraided the Zionists: "Why resist the replacement of Ladino with Greek? What use is it, to remind us of the Inquisition in Spain?"[34] After 1931, assimilationists, who pushed for greater Hellenization of the Jews, were pitted more than ever against their Zionist coreligionists, in part precisely because Greek media attacks had become more frequent and intense. Communists, socialists, Zionists, traditionalists, Venizelists, assimilationists, and religious Jews struggled to find common ground. The community was unable to present a unified front against the multiple assaults, physical and ideological, made against it.

This state of affairs is not simply a reflection of fragmentation and disunity. In many ways, it's also one of the final chapters in the history of Salonikan Jewish diversity and intellectual engagement with broader trends. Salonika's Jewish community, while treated as a corporate entity under Ottoman administrative procedure, had never been homogeneous. For centuries, different *kehilot* had struggled for primacy, and rabbis regularly argued over points of law. The city's Jews had survived the divisions brought by breakaway, heretical religious interpretation, and had debated the meaning of Ottoman reform, the implications of Zionism, and the draw of secularization. In the late 1920s, as they sought to cope with a wave of political changes, natural disasters, and demographic shifts—all within the context of a Greek administration overtly resentful of them— their multiple responses were characteristic of their traditional diversity, argumentativeness, and political and social engagement. But whereas such diversity had in earlier periods been a source of strength, in the face of the increasing external pressures of the Venizelist period it now appeared as desperation and confusion. As World War II and Nazism landed on Greece's doorstep, Salonikan Jewry was in a state of dejection and division deeper than at any time since the sixteenth century.

METAXAS TO WORLD WAR II, 1936–1941

Venizélos died, in self-imposed exile in Paris, in March 1936. The years that he ruled Greece had been hugely significant for the country as a whole and also its Jews. The 1920s marked the first-ever creation, in substantiated legal terms, of a Greek-Jewish polity—a response to the dramatic demographic shifts that the Balkan Wars brought to Greece. In large part because of these shifts, the Venizélos years were also among the most anti-Semitic in Greece's history. To a large extent, of course, the anti-Semitism of the period reflects the specific circumstances that characterized Salonika in the 1920s, and was fanned by its location on Greece's ever-volatile northern frontier. But the contradictions of the period were not unique to it. Confronted for the first time in its history with large numbers of non-Orthodox citizens, Greece under Venizélos began a process it has not yet completed: that of understanding the status of minorities within the framework of a state that embraces a specific national religion yet understands itself in liberal democratic terms. The extent to which Venizelist liberalism permitted, if not fostered, anti-Semitism is borne out, at first blush counterintuitively, by the fact that Greece under republican liberal democracy (1924–35) was far more overtly anti-Semitic than during the period that came immediately after (1936–41), when the country was led by a fascist dictator who openly admired Adolf Hitler, if not his racist ideologies. The censorship, monitoring, and social controls of the Metaxas dictatorship—designed largely to hunt out leftists—reigned in *Makedonía* and other papers, and regarded anti-Semitism as an ideology that was distasteful more for its ability to incite riot and unrest than for its specifics.[35] Greek Jews regarded him as neutral or even a supporter.[36] The result was that the final years of Greek-Jewish existence, the seven years that preceded the destruction of more than 80 percent of their number, were among their calmest ever.

Just one week before Venizélos's death, the Athens daily *I Kathimeriní* published an article titled "Tomorrow It Will Be Too Late," warning of the dangerously unsettled political situation and predicting uncertain times ahead.[37] The leftist paper *Rizospástis* published a lengthy article calling for the establishment of a "free and democratic [political] life in Greece."[38] For much of the summer, the Greek press had debated the question of whether "to Metaxify, or not to Metaxify"—was General Metaxas going to stage a takeover? One raffish editorial in *I Kathimeriní* made the wry pun that *metaxí mas*—just between us—it would be better if he didn't.[39] But by August, Greece definitely had turned in Metaxas's direction. The Left, despite its appeal to a wide array of Greek citizens and its protestations of loyal nationalism, was completely "kicked out" of the political arena, tarred as anti-Hellenist—a claim the Left itself had recently leveled at others.[40] Five months later, General Metaxas, the dictator of Greece from 1936 until his death in January 1941, established his "Fourth of August Regime." The new situation was met with a mixture of excitement, cynicism, and confusion.[41]

For the most part, Greek Jews regarded the development as a positive one. On June 26, 1936, King George and his entourage visited Salonika.

Chief Rabbi Zvi Koretz (an Ashkenazi Jew of German background) and the council of the kehila invited him to a commemorative service in the synagogue, for which Koretz composed a special prayer in Greek, asking for the health and prosperity of the king and Metaxas, and avowing the community's loyalty to the motherland.[42] No one was sure where Metaxas really stood, but felt certain that nothing could be worse than what had come before. Emphatically, Metaxas emphasized that "respect for non-Christian religious groups must be absolute."[43] He even construed some forms of Christian exclusivity as a protective measure for minority groups.[44] The Jewish press was lavish in its praise, even though Metaxas barred it from criticizing Hitler and tried to require the use of Greek rather than Ladino.[45] These measures were outweighed by his instructing the editor of *Makedonía* to stop publishing anti-Semitic articles along with his moves to disband the EEE.[46] For the most part, Metaxas simply glossed over the presence of Jews altogether. "We Greeks are for the overwhelming most part Orthodox. Of the roughly eight million Greeks who belong to the kingdom, just two hundred or at the most two hundred and fifty thousand are of other faiths, and aren't Orthodox. So all of the Greeks are Orthodox."[47]

Metaxas's racial ideology, insofar as he had one, revolved around a version of the Great Idea, and his regime was based on a bombastic vision of an ahistorical, triumphant Greece. "The Greece which was born in 1821 is not my motherland, because as a Greek I belong to that race which existed before this motherland and which belonged to another, greater, motherland. . . . I belong to that aristocracy which has already fought for its King and for the State a long time before modern Greece was born."[48] Metaxas crafted his country as "the Third Great Greek Civilization." The classical period and Byzantium were the predecessors, with the "third"— a deliberate echo of the Third Reich—to be a fusion of the two, combining the intellectual sophistication of the former with the religious piety of the latter. In this light, Jewish loyalty to Zionism was just a form of ethnic pride, no different from what he himself felt for Greece.

> Since all nations live apart from each other, every citizen should be benevolent to his own nation. I, therefore, because I was born in Greece . . . must work for the good of Greece. It is immaterial to me if Greece be the least of nations; my obligation is not to find in her virtues she does not possess, or to ignore vices that she does. My obligation to Greece . . . derive[s] . . . from the fact that I was born a Greek and . . . I therefore have to be useful to Greece.[49]

But if domestic developments weren't threatening, international ones increasingly were. The January elections had left the Communists holding the balance of power in Parliament. Following his seizure of power, Metaxas focused on squelching the Left, instilling "discipline" in the Greeks and building up the centerpiece of his ruling apparatus—the National Youth Organization (Ethnikí Orgánosis Neolaías), of which he was inordinately proud.[50] On the international front, though, Metaxas had little success. Britain shied away from signing formal alliances with Greece; only with the Italian occupation of Albania in spring 1939 were Britain and France motivated to offer both Greece and Romania a guarantee of protection should either be invaded.[51] Nevertheless, in early 1938 Metaxas wrote that he felt Greece's relations with Italy, England, and France to be *polí kalá*, "very good."[52] Reading his diaries, though, one wonders how optimistic he really felt. Throughout late 1937 and into 1938 he made frequent note of his "uneasy" and "melancholic" feelings, and fretted over his health, "watching out for the slightest problem," even though he was feeling well.[53] As he approached his sixty-seventh birthday, Metaxas suffered from nightmares and a growing sense of dread.[54] The feeling intensified as the international situation became increasingly difficult in summer 1938.

Venizélos and Benito Mussolini had signed the Treaty of Friendship, Conciliation, and Judicial Settlement in 1928.[55] Under article 28 it would be in effect for five years; if then it was not formally repudiated, it would be extended for a further five years. Within months of its expiration in late 1938, rumors were swirling that Italy had plans to invade Albania. The Greeks would find the Italians on their doorstep. The *Agence d'Athènes* published an "official Italian declaration emphasiz[ing] the fact that the rumours . . . of an impending Italian intervention in Albania are malignant falsehoods propagated with the object of disturbing the peace in the Adriatic."[56] Despite a sequence of such reassurances, by early April the region was braced for an Italian incursion. The Greek authorities urged all Greeks to stay calm and "refrain from anti-Italian demonstrations of any kind, in accordance with the desire of the Greek Government."[57] On April 6, the Italian government assured Charalambos Simopoulos, the Greek minister in London, that there were no plans to strike Albania.[58] But the next day, Skeferis, the minister in Tirana, wrote that at six o'clock that morning an Italian bombardment of Durazzo had begun.[59] A day later King Zog fled to Florina, in northern Greece.[60]

Greece suddenly had a difficult balancing act: to appease Italy, an official friend, so as not to provoke a further incursion into Greece, while at the same time appearing strong enough that Italy not think Greece would make too easy a target. At first, Metaxas was able to pull it off. Mussolini wrote to him following the Italian invasion of Albania, thanking him for his assistance in "prevent[ing] political activities on the part of Zog that might in any way jeopardize the cordial relations between Italy and Greece, the preservation of which will form the basis of my policy in the present as in the future."[61] Despite unceasing rumors that Italy's next plan was to invade Corfu, and against the manifest fears of the Greek public, Metaxas declared himself satisfied with Mussolini's guarantee.[62] For more than a year, this situation persisted: Italy issued repeated assurances that she had no designs on Greece, Metaxas reiterated his faith in the Italian promises, and the Greek public fretted over persistent rumors of a mounting Italian threat. The broader context guaranteed that this state of affairs would not be indefinite. Much as Metaxas wished to remain in Mussolini's good graces, Mussolini himself wished simultaneously to ingratiate himself with and assert himself against *his* fascist ally, Hitler.

When World War II broke out in September 1939, Metaxas wanted to remain neutral. At the same time, he wanted to maintain cordial, if neutral, relations with Great Britain, even as he sought to sustain his putative friendship with the Italians. For the first year, he managed to maintain this delicate balance, at the cost of ignoring the increasing evidence that contrary to Mussolini's assurances, Italy did not plan to honor its promises of respect for Greek territorial integrity. Metaxas spent the early months of 1940 enjoying the cold weather, playing pinochle, and sleeping in. He spent the day of March 17 in bed, pondering Greece's situation vis-à-vis Mussolini and Hitler, and concluding that there would be peace.[63] He was wrong. Eager to show Hitler that the Italians could mount successful military campaigns on their own, Mussolini turned to Greece as an easy target. In August, the Italians torpedoed the Greek warship *Elli* in the harbor at Tinos, on the day of the Feast of the Assumption of the Virgin, one of the most important religious holidays of the Orthodox calendar. Metaxas, tears in his eyes, monitored the situation but maintained his staunch neutrality.[64] The papers predicted that soon Greece would be at war.[65] "An Italian attack seemed imminent. But the summer passed and nothing occurred."[66] Soon enough, though, less than three months later, predictions proved correct. At three o'clock in the morning of October 28, 1940, the Italian minister in Athens was dispatched to Metaxas's home, bearing an ultimatum that Italian troops be allowed to occupy Greece. The letter stated that Greece could not be trusted because of the British guarantee of her sovereignty, and Greek permissiveness regarding British use of Greek air bases in Thessaly and Macedonia, and ports and harbors for naval operations. Construing his demand as a "guarantee alike of the neutrality of Greece and the security of Italy," Mussolini's ultimatum to Metaxas concluded: "The Italian Government demand that the Hellenic Government instantly issue the necessary orders to the military authorities, so that the occupation may be carried out peacefully." In the event of any resistance from the Greeks, Greece "will be crushed by force of arms, and in that case the Hellenic Government will bear the responsibility for whatever will ensue."[67]

Fig. 6.1. Members of the "Koen Battalion," 1940, Salonika, Greece. Permission Kehila Kedosha Janina, New York.

Famously, Metaxas's reply was an unqualified "no," *óhi*. He elaborated on the response in a series of missives written later in the day to his minister for foreign affairs, the royal legations abroad, the minister in Rome, and the Greek people.

> Italy, denying to us the right to live the life of free Hellenes, has demanded . . . the surrender of portions of the national territory. . . . I replied to the Italian Minister that I considered . . . the demand . . . as a declaration of war on the part of Italy against Greece.[68]

The attentions of Greek Jews were focused on Greece and its defense. After the flurry of migrations to Palestine in the late 1930s, now virtually none thought to leave the country, even in the face of a war that clearly threatened to bring Nazism to Greece's doorstep. In 1940, more that five thousand Jewish refugees from other parts of Europe passed through Greece en route to Palestine. The American Jewish Joint Distribution Committee worked with Greek Jews in Athens to raise money on their behalf and send them on their way.[69] While some of the refugees remained in Greece, almost no Greek Jews left that year for Palestine; in fact, in the seven years before the Italian invasion, less than 1 percent of Jewish immigration to Palestine was from Greece.[70]

In the rapid mobilization of troops, thousands of Jews were conscripted.[71] The Fiftieth Regiment (the so-called Koen Battalion), from Salonika, was mainly Jewish, and one of the first and most famous high-ranking Greek officers killed in the defensive campaign, Mordechai Frizis, was a Jew.[72] Among the soldiers who served with the greatest distinction were prominent Jews from Jannina and Salonika. Jewish veterans who survived the war would later be taken from the hospitals where they were recuperating and sent to Birkenau by the Germans.[73] Others would return to their homes to discover Nazi policy in full effect. Itzchak Nechama, a Salonikan who fought on the Tepelene front, was released from service in March 1941, in Jannina. After a twenty-eight-day walk, Nechama reached Salonika only to find that many of his possessions had been confiscated by the German authorities.[74]

Metaxas's belief that he might be able to contain the conflict, limiting it to a matter best settled between Greeks and Italians, led him to decline Winston Churchill's offer of assistance, thinking that by doing so he could keep Hitler and the Germans out of it. He did, however, accept British munitions—antitank rifles flown in from Egypt.[75] The Italians, instantly and unanimously hated, became the subject of mocking cartoons and satiric pieces in the Greek press, and Mussolini, depicted as a bumbling bear, was universally ridiculed.[76]

For their part, the Italians had evidently had Greece in their sights for quite some time. They also had not necessarily expected Greece to agree to Mussolini's ultimatum. The Italians were ready to fight for Greek compliance, if needed. On October 26, two days before the ultimatum was issued, Licurgo Zannini, the commander of the "Ferrara" mountain infantry division, had written to his troops, "For nineteen months we have been sharpening our weapons and hearts, in this well-defended and rugged country of Albania, straining toward the goal which is now in sight."[77] At a meeting on October 15, 1940, at the Palazzo Venezia, Mussolini had revealed that invading Greece had been on his mind even longer than that: "I have been pondering on this action for long, for months and months: before our participation in this war, even before the beginning of the conflict."[78] Metaxas's view of Italian intentions had clearly been optimistic, but he is remembered to this day for his categorical refusal to capitulate to Italian demands rather than his stubborn naïveté.

During the first weeks of battle, Metaxas knew he was living through the most important moment of his career. In his diary, he began measuring the date according to the number of days Greece had been at war. His moment of glory was short-lived. Within three months, Metaxas was

Fig. 6.2. Mordechai Frizis and his wife, Victoria Costi. Permission Photographic
Archive, Jewish Museum of Greece.

dead, after eighty-two days as leader of Greece at war. His successor, Alexandros Koryzis, reversed official policy, welcoming a British expeditionary force to the region in the hopes of ending the deadlock that had prevailed through winter 1940–41. This move brought Greece out of the
regional arena and into the international one. But ineffective communication between the British and the Greeks led to tactical errors in western
Macedonia.[79] On April 6, 1941, the Germans swept through Yugoslavia
and Bulgaria.

"Salonika was now at the mercy of the invader. . . . [T]he last train
would leave that afternoon—the 8th—at three o'clock. Simultaneously
. . . sappers began to carry out the prearranged demolitions of bridges,
roads, and railways that had been prepared in the event of Salonika falling
into German hands."[80] Within days, Salonika was isolated and under German occupation. By April 23, the Germans had made their way to Athens.

Only during the three short decades spanning the Balkan Wars and World
War II was Greece a significant Jewish European center, home—by dint of
its conquest of Salonika—to one of the most distinguished and confident
Jewish cities on the continent. Over the course of thirty years a fledgling
Greek Jewish identity began tentatively to emerge. Before the brutal inter-
ruption of the Holocaust, a whole generation of Salonikan Jews would
come of age as Greeks. Testimonials from Salonikan Holocaust survivors,
gathered in vast majority from men who were in their twenties and thirties
during the war, reveal the extent to which this generation regarded Greece
not simply as the place they lived but as their patrída, their fatherland,
and the extent to which they considered themselves Greek.

Considered in context, this development was nothing short of remarkable. It suggests that two narratives that have long dominated the historiography of Salonikan Jewry need some revision. The first is that of Salonikan Jewish insularity, and the second that of 1920s Greek xenophobia.
To be sure, both narratives reflect certain irrefutable realities: Salonikan
Jewry was renowned for its sense of superiority and apartness, both vis-à-
vis other Jewish communities and Greek Orthodox Christians; Salonika's
transition from an Ottoman to a Greek city was punctuated by anti-Semitic outbursts perpetrated at both the state and personal levels. But this
is only part of the story. The three brief decades during which Greece was
one of the premier Jewish nations of Europe also saw the maturation of
a generation of Greek Jews who understood themselves not as a paradox
but in relatively uncomplicated terms. That this was the case signals the
extent to which Salonikan Jewish society's resilience relied as much on
flexibility in the face of change as the ability to resist it. To be sure, these
developments were as much the inventions of necessity as anything; the
extraordinary pressures that a century of Greek territorial expansion

Fig. 6.3. German army invasion of Greece, April 7, 1941. Permission Film and Photo Archive, Yad Vashem, Jerusalem.

placed on the region's Jews, combined with the social, economic, and political traumas that came in staccato succession—1897, 1912, 1917, 1923, and 1931—inevitably prompted change of an unwanted sort. Assimilation, secularization, and Hellenization were largely accommodationist responses to the changed circumstances that Greek Jews found themselves in by the interwar period. But they were not only that. The circumstances of life under Greek rule also presented the possibility for a new, more modern and integrated Jewish existence. It is unclear whether or not a full-blown Greek Jewish culture, given time, might have emerged; it likely would have. Yet to imagine such a thing is to imagine away everything that followed.

Chapter 7

Occupation and Deportation: 1941–44

All these people are dead now, *malheureusement!*
—Cited in Moise Pessah, *Faces and Facets*

A PHOTOGRAPH from turn-of-the-century Crete shows a family gathered in their garden for their portrait.[1] They all sit somberly for the occasion, with the exception of the two young men in the upper left corner of the frame who mug for the camera, posing coquettishly behind a rosebush. Most of the men have bushy handlebar mustaches, in the Cretan style, and they hold Cretan instruments—bouzoukis and a fiddle. All wear modern dress, save for the patriarch, who sports a fez. In the front, kneeling in the dirt, are two small children, a boy and a girl. These children's parents were of the last generation of Cretan Jews who would have the luxury of dying of natural causes. Starting May 20, 1944, the entire Cretan Jewish community—278 people—were rounded up, kept for two weeks, then sent by boat from Hania to Heraklion, and locked in the Venetian fort.[2] The German officers who pounded on their doors at the break of dawn told them they had twenty minutes to get ready to leave. Orthodox schoolchildren arrived in school the next morning to discover their classmates gone. On Crete, where the tiny Jewish community was highly integrated into the dominant Orthodox society, the Christians were given no time to help, and the Jews no time to escape. In Heraklion Crete's Jews were put on the ship *Danae* for the first leg of the journey to Auschwitz. On the same day that the Allied forces were arriving in Normandy— June 6—the boat set sail. On board along with Crete's Jews were several hundred Greek hostages, along with hundreds of Italian soldiers. Their ship was torpedoed and sunk off the coast of Folegandros within two days of its departure, likely by a British torpedo. No one on board survived.[3] On the day after the roundup, a Christian Cretan schoolgirl recorded her sorrow. "Oh, they were Greeks, they were our brothers, and their children had fought alongside us for freedom."[4]

On Zákinthos (Zante), off the western coast of the Peloponnese, all 275 Jews survived. When the island's mayor, Loukas Karrieras (Karrer), was ordered—at gunpoint—to provide the occupying Germans with a list

Fig. 7.1. Jewish family, ca. 1900, Hania, Crete. Permission Photographic Archive, Jewish Museum of Greece.

of all the island's Jews, he and metropolitan bishop Hrysostomos turned over only two names: Hrysostomos and Karrer.[5] If the Jews were to be deported, the bishop is said to have maintained, then they, too, would have to go.[6] While Hrysostomos and Karrer argued with the Germans, all the Jews were taken and hidden in the mountains with Christian families. The mayor, the bishop, the chief of police, and every member of the Christian population worked together to hide the island's Jews. Only the old and the sick remained behind in the island's main town. The SS planned to have them picked up by a ship en route deporting Jews from Corfu, but ultimately did not do so.[7] No ship came to the island for them, no one was deported, and all survived the war. After the war, most Zákinthiote Jewish families moved to Palestine. When the island was devastated by an earthquake nine years later, among the first boats to arrive with aid was one from Israel, with a message that read, "The Jews of Zákinthos have never forgotten their mayor or their beloved bishop and what they did for us."[8] In 1978, Karrer and Hrysostomos were honored by Yad Vashem as "righteous gentiles."[9] The collective memory of the salvation of these Jews has ongoing importance to the island, which over the decades has held numerous ceremonies in commemoration.[10] "Zákinthos

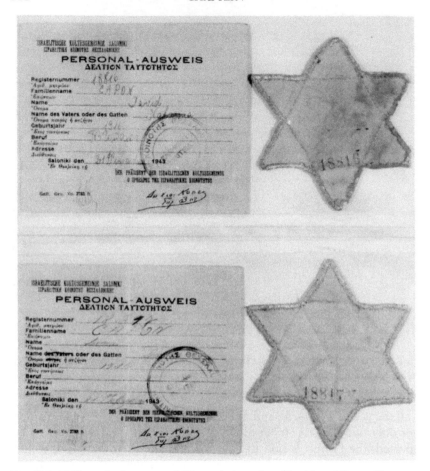

Fig. 7.2. Yellow cloth stars and ID cards issued to David and Luna Capon on February 21, 1943, in Salonika, in order to identify them as Jews. Permission Photographic Archive, Jewish Museum of Greece.

feels proud that it managed to carry the torch [of honor and virtue] from the narrow boundaries of its local history to the wider limits of our national [Greek] history."[11] While virtually no Jews have lived on Zákinthos for over fifty years, Zákithiotes and their descendants living abroad still regard the island as their patrída, or homeland.[12]

Not too many miles north, on the island of Corfu, Mayor Kollas collaborated with the Germans, facilitating the deportation and murder of well over three-quarters of the island's two thousand Jews. Incredibly, even though the local German territorial commander, Oberst Emil Jaeger, actually worked to prevent their deportation, and despite some sympathy for the Jews among the local Greek population, only two hundred survived.[13] Rounded up on June 6, 1944, the deportees were sent in two shifts by boat to Patras, and from there overland to the Haidari concentration camp in Athens, where they were put on transports to Auschwitz. Corfu's Jews had the dubious honor of constituting the last Jewish transport sent to Auschwitz from Greece.[14] Kollas issued a public statement congratulating the Germans on their success, thanking them for ridding Corfu of this "foreign element" that had taken over the island's economy, and declaring that all property on the island had reverted to the Greeks.[15] When 185 Jews tried to reclaim their homes after the war, only 61 of them were successful.[16]

Two of the roughly fifty thousand Salonikan Jews who died in the concentration camps were the husband and wife, David and Luna Capon. They were issued identification cards and yellow felt stars, identity numbers 18816 and 18817, in February 1943.[17] They were deported along with the rest of their community in the spring of that year. The transports from Salonika, which took place between March and August 1943, were the first to leave Greece. The vast majority of the Jews who arrived on the early Greek transports to Auschwitz were gassed on arrival. David's brother Nehama was one of the tiny minority who survived.

In each case, the fate of Greek Jews was determined by the local authorities, cultural factors, timing, and the impact of specific individuals or events. Differing regional survival rates reflect the range of attitudes toward Jews that prevailed in different parts of the country as well as of Jews toward Greece. The emergence of Greek nationalism among Jews, first during Metaxas's rule and then under German occupation, led many to participate in the Greek resistance, which was the only Greek wartime institution with even a semiofficial policy of helping Jews avoid deportation. Steven Bowman has recently documented the scope of Jewish participation in the resistance, and the impact it had on Jews during and after the war.[18] Hellenized Jews—those in Salonika with contacts around the country, and particularly those in Athens—had higher survival rates than did the core Salonikan community. As a generalization, smaller communities fared better than large ones. Just as Jewish life in the south Balkans had for centuries been conditioned as much by local circumstance as by formal policy, the degree of devastation brought by the Holocaust varied from locale to locale according to a complex set of circumstances.

SALONIKA, APRIL 1941–AUGUST 1943

The deportation to Poland and elsewhere of the entire
Jewish population of Salonika . . . [was] an
unparalleled mischief in the annals of human history.
—I. Nefussy to British Embassy, Athens,
January 20, 1945

At roughly the same time that Metaxas was passing a difficult winter in Athens, racked with physical aches and pains and filled with a psychic sense of impending doom, German diplomats in Greece were given their first orders to begin gathering information on the country's Jews. The German missions in Salonika and Athens began amassing information on the size of the local Jewish population, its characteristics, and the degree to which it was assimilated into Greek society. By the time German troops first arrived in Salonika on April 9, 1941, the Third Reich had acquired a good deal of basic information.

The recent history of Greek Salonikan anti-Jewish activity attracted particular German interest. On their arrival in the city, the Germans began systematically to exploit intercommunal tensions even as they set about gathering material artifacts of Greek Jewish life. Within two weeks, the "Rosenberg Sonderkommando," a special task force dedicated to gathering material artifacts for Frankfurt's new Library for the Exploration of the Jewish Question, was on the scene. Johannes Pohl, director of the Department of Hebrew at the Institute for the Study of Judaism, led the search for valuable documents.[19] The city's rabbinic leadership was completely demoralized by the loss of its sacred texts. "One day three trucks arrived and took the complete library of old books which we had in the Community, books from before the time the Jews left Spain. Our Rabbis cried so much when they took those books, I remember it to this day. I recall the tears of these learned men who said: Nothing matters to us, only these books."[20]

The Germans reestablished the EEE, shut down Jewish newspapers as well as religious and civic organizations, and encouraged a boycott of Jewish businesses.[21] By summer 1941, a completely new way of life had set in. From the start of the occupation, Greece's Jews were at the center of Germany's plans for the country. The consolidation of German control over northern Greece and the construction of much-needed transportation links were no more important than the systematic isolation of the Jews.

Some Jews left the city right away, but not many. For those who had business contacts or relatives elsewhere in Greece, particularly in Athens, this was an easier proposition than for others. Elia Aelion helped move his family's business to Athens in the weeks after the German arrival. The family had the business shipped in installments, to different contacts, all Christian, who hid the goods. Aelion also sought out a safe house in Athens for each of his relatives back in Salonika, should they need to leave as well, although in the end it was impossible to get them out. Once in Athens, "In contrast to my experience in Salonika, I felt safe as a Jew and could move about freely."[22]

Many of the city's Jews held European citizenship, most commonly Spanish or Italian. Some sought out the assistance of the local consulates, although by autumn 1942, only the Italian consulate remained open.[23] The Spanish government was extraordinarily parsimonious in its recognition of Sephardic claims to Spanish nationality both during the occupation and after.[24] In contrast, the Italians consistently insisted that they had the obligation to protect the Italian citizens, regardless of race, who were living under German control. A series of memorandums in late October 1941 show that worries about what the Germans might do to Salonika's roughly three hundred Italian Jews were already in the forefront of the Italians' minds. Consul General Pietro Vitelleschi received a letter from a delegation of Italian Jewish Salonikans who wrote of their "devozione alla Madre Patria" [devotion to the motherland].[25] Vitelleschi gave assurances that the Italian government would extend them all possible protection. Its ability to do so was greatly impeded, however, by the fact that in June the country had been divided into three different Axis zones of occupation. The Italians held the mainland, the Dodecanese, and eastern Crete; the Bulgarians controlled most of Thrace and parts of eastern Macedonia. This left the Germans in control of the most heavily Jewish areas—Salonika and the Macedonian hinterland—along with most of Crete, eastern Thrace, and the northeasternmost Aegean islands. Athens was coadministered by the Germans and the Italians. Italian interference in occurrences outside the Italian zone was not welcome; in summer 1942, Consul General Guelfo Zamboni noted that the Germans had expressed unhappiness at Italian attempts to intervene in goings-on in Salonika and elsewhere.[26]

Salonika's Jews were not deported until spring 1943. Life under the Germans before that was far from easy, though. By September 1942, the city was in a severe state of famine.[27] Everything that the region's farmers produced—wheat, butter, cheese, oil, olives, fruits, and nuts—was stolen by the Germans. Even the fish that Salonika's commercial fishers pulled from the sea were taken away.[28] What the occupying forces did not need was sent to Germany. During one year of the occupation alone, 520,000 tons of food and other commodities were shipped out of Salonika to Germany.[29] Suspected leftists, many Jews among them, were shot and hung by the score by the German authorities.[30] But the Greeks hung on. The Germans and the Italians alike had underestimated their powers of resistance from the start, first in Albania and now as the occupation set in. A British lieutenant colonel on the scene marveled at the tenacity of the Greeks and the Germans' failure to anticipate it. "Poor deluded Teuton, when will you realize that in the Mediterranean you are dealing with men who are intelligent and realistic?"[31] Later, in Auschwitz, this same depiction of the resourceful Mediterranean type would be applied by Ashkenazic Jews to the Greeks they met there.

As time went by, the situation worsened. The excruciating circumstances in Greece, particularly the widespread famine, were held up by

President Franklin Delano Roosevelt as the ultimate example of Axis cruelty. In 1943, two years into the occupation, he wrote, "Today Greece is a land of desolation, stripped bare of all the essentials of living. Thousands have died of hunger. Thousands are dying still. Greece today is a gaunt and haggard sample of what the Axis is so willing, so eager to hand to all the world."[32] By late 1942, the outside world was aware that Greece's Jews were being subject to especially harsh treatment: forced labor and imprisonment.[33] When Greece's Jews were deported to Auschwitz-Birkenau, they were particularly ill prepared to survive the privations of the transports and the camps. After starving in Greece for two years, they arrived in Poland emaciated; 80 to 90 percent of them were sent directly to the gas chambers. "When the anti-Jewish measures started, the Jews were in such a state of undernourishment that even Hoss and his helpers did not know what to do with them when they arrived in Auschwitz."[34] During the occupation, Greek Jews along with Asia Minor Greeks fell victim to famine in disproportionately high numbers; these groups tended to be members of the "unemployed proletariat," with little access to ready cash and few contacts in the hinterlands.[35]

In Salonika, the horrors of the occupation furthered the effects of the city's long series of catastrophes: Jews and Christians were pushed further and further apart, and the long-standing rivalries between them came into full bloom. The Germans encouraged these divisions, with the German-controlled Greek press, particularly the papers *Néa Evrópi* [New Europe, the Nazi mouthpiece *Neue Europa*] and *Makedonía*, publishing articles outlining supposed Jewish crimes against the Orthodox Greeks. Venizelist Jews, members of the Liberal Party, tended to have more Orthodox friends and were more likely to find assistance fleeing southward, to the Italian-held parts of the country. The wealthy also found flight from the city relatively more easy than most. But for everyone, leaving was not a ready or simple option, and most stayed put.[36]

In summer 1942, the situation for Salonika's Jews rapidly deteriorated. On Saturday, July 11, German authorities ordered all Jewish men between the ages of eighteen and forty-five to Eleftherfa (Freedom) Square.[37] The square became "black with people. Nine thousand adults were massed there, under a sun of lead, and were made to stand erect, in rows, without moving, beneath the eyes of German soldiers, from eight in the morning until two in the afternoon." Anyone who tried to shield themselves from the sun, under a hat or a newspaper, or behind dark glasses, was severely beaten. The men were forced to roll on the ground and perform calisthenics "of the most burlesque sort." Those who collapsed from heat or exhaustion were doused with water and kicked until they stood back up.[38] The tortures were led by Germans from all units—army and navy as well as SS. Visiting German actresses stood on balconies above the square,

Fig. 7.3. Roundup of Jewish men, July 11, 1942, Eleftheria Square, Salonika. Permission Photographic Archive, Jewish Museum of Greece.

taking photographs and applauding.[39] Finally, the men were released. "The SS ordered the victims to go back to their homes and required them to run the first 150 meters or to go on all fours, turning somersaults or rolling in the dust."[40]

During the event, many Greek bystanders appeared either indifferent or amused.[41] *Néa Evrópi* used the occasion to fan the flames of anti-Semitism. The next day, the paper published photographs of the public humiliation in a front-page article, writing that "non-Jewish spectators, gathered in the surrounding road . . . had but one wish: that scenes such as the one they'd just seen would go on as long as possible."[42] It is unclear what most Christians actually made of the spectacle; it is likely that many felt horror and pity. Michael Molho reports that when, some months later, Jews were obliged to wear the yellow star, many Christians regarded them with sympathy and took "the treatment to which the Jews were subjected as proof of the barbarity of the Germans."[43] Nevertheless, during the long course of the day, no Greek authority interceded on the Jews' behalf.[44] On the whole, it is unclear to what extent Greek authorities assisted the newly installed German administration. The May 11, 1942, edition of the Salonikan resistance (the EAM) newspaper *Eleftheria* included a sidebar

article urging local Greek police not to collude with the Germans, re-
minding them that "you, too, are brothers of the same Greek people."[45]
This same spirit of brotherhood seems not to have extended itself to the
Jews, whose plight for the most part was—from a combination of fear,
hopelessness, and callousness—ignored, at least on an institutional level.

Days later, the Jews were again summoned to Elefthería Square; now,
all men between eighteen and forty-five were sent as forced labor to various parts of Macedonia, where they worked in mines and quarries, on
roadworks, and building airports, helping the Nazis rebuild the infrastructure so recently destroyed by the retreating Greek forces. Only those
Jews who held Italian or Spanish citizenship were exempted.[46] The workers were given little water and only three hundred grams of bread per day
as rations, along with some soup made of sour cabbage.[47] Tuberculosis,
typhus, malaria, and dysentery killed hundreds of them.[48] Others were
beaten or kicked by German guards until they died.[49] As a result of much
lobbying, early that fall the community elders were able to gain the release
of just over one thousand, at the cost of one million drachmas per person.
Again, no official Greek government or church intervention was made.[50]
Negotiations instead were between Jewish community leaders and the
Germans. In late October, when the cold weather would in any case soon
force a halt to the labor, the Germans approached the Salonikan chief
rabbi, Koretz, and told him that for a cost of several billions of drachmas
he could have the rest of the workers freed as well. A further condition
appears to have been the demand that the Jews turn over their cemetery.[51]
Koretz, who had already handed over the community's registers to the
Germans, continued to comply with German demands; whether he did so
out of collusion or the naive belief that compliance would bring leniency
is unclear.[52] (Koretz had been unpopular with the community from the
outset, as he was a non-Sephardic, German modern Orthodox rabbi and
perceived as an outsider.) In any case, during the first week of November,
about seven thousand remaining workers returned, at a cost to the Jewish
community of 3.5 billion drachmas.[53] Years later, in the Adolf Eichmann
trial, Koretz was branded a traitor for his "criminal cowardice" for acced-
ing to various German demands.[54]

By this time, about twelve hundred Jews, mainly the wealthy, had fled
the city for the Italian zone.[55] At the time, it cost 150,000 drachmas—
about £300 sterling—to hire a car to Athens.[56] Even for those who could
afford it, the decision to leave was excruciating. Many who considered
leaving decided not to, on account of the family members they would
have to leave behind.[57] Those who were more inclined to flee—the young
and healthy—felt that they should stay behind to care for their parents.
"[Christian friends] told us [to leave], but, you know, there was a lot of
love between family members. Families didn't split up. They said, where

our parents go, we're going too. We thought that we could save them. Because we were young. We'd be able to do something for them."[58]

That month, Zamboni wrote to the Italian Diplomatic Mission in Athens that in light of the worsening situation, the Italians should arrange for the transfer of needy Jewish families to the Italian zone and move to protect any property that they might leave behind.[59] In August, Italian efforts to assist Greek Jews both in Greece and abroad—notably in Bulgaria and Belgium—had intensified, but soon Rome advised its consulates that "it would be wise . . . to suspend the initiatives taken on behalf of Jews of Greek citizenship."[60]

During winter 1942–43, the full contours of Nazi anti-Jewish policy came into stark definition. On December 6, 1942, the Germans began to dismantle the Jewish cemetery, sending five hundred Greek workers to the site, where they uprooted about a hundred thousand tombstones. Some Jews were allowed to remove their relatives' bones, but the vast bulk of the skeletal remains were strewn across the area. German soldiers picked through the bones, looking for gold teeth, jewelry, and anything else of value.[61] For religious reasons above all, of course, the desecration of graves was unspeakably traumatic. But the destruction of the cemetery also marked the final diminution of Jewish physical presence in the city. After the great fire and the influx of Asia Minor refugees, the cemetery had remained just about the only definitively Jewish architectural feature of the landscape, and had long been a popular gathering site for relaxation and recreation. The cemetery, more than four hundred years old, was immense, and with the rapid growth of the city after the turn of the century it had become a target for urban expansion. Conflicts between Christians and Jews over the territory as early as 1925 had been resolved, albeit warily, in favor of the Jewish community. The Greek Christian workers brought now to dismantle the cemetery appear to have been only too happy to help.[62]

Throughout the previous summer, *Néa Evrópi* had called for the implementation of racial laws and published editorials arguing that Jews should have to wear some sort of distinguishing mark on their clothing. These sentiments were fulfilled in February 1943, when Dieter Wisliceny of the Sicherheitsdienst (security police) was sent by Eichmann, who had visited the city in summer 1942, to oversee preparations for the deportation of the city's Jews. Wisliceny's commission, "made up of six real vampires," immediately set about its business.[63] Within two days of their arrival, they summoned Koretz to inform him that the Nuremberg laws were to be imposed on the city.

Jews were to wear a yellow star, move into the "Baron Hirsch" ghetto, obey a curfew, and cease all interactions with non-Jews.[64] Any non-Jew who entered the ghetto was to be shot on sight.[65] Many, however, did—

Fig. 7.4. Above, the site of the ruined Jewish cemetery of Salonika, with its five-hundred-year-old tombstones, broken into pieces. In the foreground, several of the tombstones have been used to line a swimming pool for the German forces. Permission Photographic Archive, Jewish Museum of Greece. Below, the ruins of the Jewish cemetery, 1945, Salonika. Permission Film and Photo Archive, Yad Vashem, Jerusalem.

at least before the Germans became strict in their enforcement of the policy. "How thrilled I felt when my Christian friends came to visit us in the ghetto," recalled Erika Amariglio. "All of our friends came to see us. Some of them came every day!"[66] An official order, signed by Max Merten, the head of the city's military administration and dated the same day as Wisliceny's arrival, February 6, outlined the new situation. The Jews had until February 25 to comply. A flurry of further orders, issued over the course of the next week, stipulated that all Jewish businesses and homes must be marked prominently as such, described in detail the gold star that was to be worn, defined racially who counted as a Jew, and forbade Jews from using the telephone.[67] Pianos, a hot item among the occupying Germans, were also confiscated. The Germans wandered into Jewish stores, helping themselves to whatever they fancied, without making any payment for it.[68]

The Italians tried to make good on their earlier promises; Pellegrino Ghigi, the minister plenipotentiary for Italy in Greece, had over the course of the previous summer written of his objections to these policies, particularly the requirement that all Jews wear the yellow star, but ultimately to no avail.[69] In any case, by the time the laws went into effect in Salonika, the Italians had little time to mobilize against them.

Wisliceny, in conjunction with the SS, oversaw the roundup of the community into the ghetto. On February 26, 1943, the German consul general Fritz Schönberg wrote to the German Foreign Ministry, reporting that the fifty-six thousand Salonikan Jews registered with the authorities were being concentrated in Baron Hirsch.[70] Bands of Jewish police came nightly to the Jewish homes in the city, gathering up women, children, and the elderly in carts to transport them to ghetto. The rest walked. "The roads were full of big wagons, little wagons, bicycles loaded with mattresses, beds, chairs, suitcases and bundles. Men were pushing, children young and old helped as best they could, and women were following, babies in their arms or dragging their children by the hand."[71] German soldiers stood by waiting for the people to leave; as soon as they did, the soldiers descended on their homes and took anything of value that had been left behind.

In the ghetto, everyone was in a panic. Anyone of marriageable age rushed to find a spouse. In the first weeks, up to ten marriages a day took place. Soon people were marrying by the hundred—the boys so that they could find connubial comfort, and the girls in the hopes that being married might give them some protection. When deportation began, married couples also "had the privilege" of going on transports together, so marriage was a way of choosing with whom one would pass the awful journey.[72] Two of Iakovos Hantali's brothers married in the ghetto, as did his sister. "I don't remember her husband's name," he later wrote. "I only remember his face. There wasn't time for niceties. . . . Within one week we had three weddings in my family."[73] Funerals were common as well, as "the number of those in Thessaloniki committing suicide in their despair rose by the hour."[74]

The next month, Koretz was summoned to hear even worse news. The city's Jews were to be deported.[75] As of March 6, Jews were no longer allowed to move about the city. An unarmed police force of two hundred Jews was established to ensure that order in the Jewish quarter was maintained. German officers told the Italians that fifty-one thousand Jews were to be taken on a forced March to Poland.[76]

The sudden confinement of the Jews to the ghetto had a devastating effect on the city as a whole. Its impact gives a clear picture of the absolutely integral role Salonika's Jews still played in the city's economic life even in early 1943. And the Greek reaction to it suggests that poor Christian-Jewish relations in the city stemmed largely from economic factors, and were not as virulently anti-Semitic as *Néa Evrópi* hoped:

> The sudden disconnection from the life of the city of more than fifty thousand people who for centuries had undertaken important activities there, has created a most difficult state of affairs. All of the services that had been provided exclusively by the Jews are now missing, and prices have gone through the roof, so much so that the [remaining] locals have turned against the Greek merchants who have benefited from the situation. . . . The Greeks, aside from the merchants who have benefited from the lack of competition, show no enthusiasm for all that has been done [to the Jews].[77]

Within days, the transports had begun.[78] Jews were to be deported at the rate of ten thousand per week, with sixty people packed into each cattle car. The Italians fruitlessly suggested that the Jews be concentrated on one of the Greek islands instead, and the attempts at intervention by the Red Cross were rebuffed.[79]

Simultaneously, deportations began from other towns in the German zone. Veroia, southwest of Salonika, was home to about six hundred Jews and had close ties to Salonika. Relations there were good between Jews and Christians. The synagogue's valuables were hidden with the Orthodox metropolitan bishop Polycarp, who kept them in his own home until after the war. When the president of the Jewish community, Menachem Stroumsa, learned in a phone call with Jewish leaders in Salonika that deportations had begun, he "urged the people to abandon Veroia. To hide themselves wherever. But it's difficult for anyone to leave everything, all at once; what had been our life and our small holdings." Fifty of the town's Jews fled to the mountains with the help of Greek Christians; 550 were sent to Auschwitz, soon after Florina's 500 Jews had been taken away.[80] The community's rabbi, Azaria, was among those who hid with Christians during the war. "We knew that a harsh fate waited in Poland, but nobody could imagine the horrors of the gas chambers. We didn't learn until after the war. Twenty centuries old, our community was lost in a few days."[81]

The news of the deportations rapidly reached western Europe. In June, the London Jewish press reported that three thousand Salonikan families had been sent to Poland.[82] By January 1944, it was known that the deportations were more or less complete.[83] There is debate over how much the Salonikans themselves knew about their fate. It is likely that by early 1943, many were sure that they were to be deported from the city. It is equally likely they thought that at worst, they would end up doing hard labor once they arrived in Poland. In January 1943, the Italian consul in Salonika had written to Athens about growing fears that deportation was near. "During these days persistent voices have been circulating, the basis of which is unknown, about measures soon to be taken against the Jews of Salonika. There is talk of deportation to Poland. . . . Naturally this has produced great agitation among the large Jewish community."[84] On February 27, *Néa Evrópi* had published an article calling for Jewish deportation.[85] The Jews reacted with panic, and according to Italian eyewitness reports the Greeks expressed no support whatsoever for the notion.[86]

The implications of the precipitous rise in anti-Semitic legislation and specifically anti-Semitic measures, relentless since summer 1942, must have led many to wonder what the ultimate outcome would be. As the deportations got under way, those left in the ghetto became gradually more aware of the horrors that awaited them. Some of the transports were driven by Greeks as far as Belgrade; drivers returned to Salonika shocked by the harrowing sights they had seen on the way.[87] A trainload of Jews sent from Didimoticho to Salonika for deportation to Auschwitz stopped along the way. Some Greek Christians who encountered the train on one of its stops were horrified by what they saw, and spent hours going from car to car with water, bread, cheese, and candles until the German guards intervened and sent them away.[88] The reports of such eyewitnesses reached many of the Jews who had yet to be deported. The train ride to Poland took close to three weeks, during which the passengers could not sit down and had nothing but a tiny bit of bread to eat. In some wagons, the death rate on the journey was as high as 50 percent. Several people went mad along the way, and virtually no one was able to stand up on arrival.[89]

As soon as the transports began, the ghetto was overrun with chaos. Greek criminals ("bande di malviventi greci," as the Italians called them) rushed in to steal any possessions the deportees had left behind, and members of the Greek resistance dispatched messengers who called on the Jews to flee to the mountains and join them.[90] Although most did not want to leave their families, some did so.[91] Others fled for the Italian zone.[92] But the vast majority was unable to go anywhere at all. A Jewish militia of two hundred Polish, German, and Greek Jews patrolled the camp, torturing wealthy detainees to learn where they had hidden their valuables and forcing young women to work as sex slaves.[93]

Throughout March and April the transports continued, slowing slightly only when the Bulgarians turned over all the Jews in their occupied zone en masse, and some trains were diverted for their deportation.[94] The Italians, who thought these Jews were being sent to work in a Warsaw factory, were shocked by the Bulgarian compliance with German anti-Jewish policies and expressed happiness at the fact that some thousand Jews in the Bulgarian zone had managed to escape, mostly to Albania.[95]

As the ghetto emptied out, over the course of April and May, the Italian consul Zamboni struggled valiantly to stop the deportations, even going so far as to send a secret message to the consulate in Athens that the Germans had plans to send police to the city to search for Salonikan Jews who had escaped.[96] The next month, when several Jewish police were in fact sent from Salonika to Athens to look for fugitives, Zamboni wrote again to Athens to warn them.[97] In Athens, the arrival of these spies caused many Salonikans who had fled to quickly relocate to the suburbs.[98] By the end of April, the Germans were openly accusing the Italians of smuggling Greek Jews out of the city.[99] At the beginning of May, the Italian Consulate in Salonika was issuing certificates of Italian citizenship to Jewish Greek nationals.[100] When the Italians learned from the German authorities that several Jews with Italian citizenship had already been deported, they demanded their return.[101] Zamboni also asserted the right of the Italian government to deal with the case of Katerina Salunia, a Christian of Italian citizenship who had been charged with giving aid to Jews.[102]

By the middle of June, almost all the Jews had been deported from Salonika. Among the few who had been left behind were the young girls taken off the transports at the last minute by German guards, who sent them to work in military brothels.[103] Most Jews had been sent by the June 15 deadline established by the Germans. Italian Jews were given a month's extension; they had until July 15 to leave the city or be sent to Poland. Just after midnight on the morning of July 15, a train of 331 Italian Jews left for Athens after fourteen hours of negotiations between the Italian consul Zamboni and German authorities.[104] Nine days later, the Italians arranged for nine Argentinian Jews to be sent to Athens as well.[105] The Italians also tried to secure the safety of Spanish citizens, but when Spain refused to allow for their return to that country, they were sent with the transports to Poland.[106] By the beginning of August, the ghetto was a ghost town. On the night of August 10–11, the last transport departed.

Hours later, having finished its work, the special SS unit left the city, and Wisliceny flew to Berlin. On the morning of August 11, Consul General Castruccio, surveying the empty streets of Salonika, wrote: "The Jewish community of Salonika, which was founded before the discovery of America and which included around 60,000 members, exists no more. . . . The liquidation of the Jewish community was carried out with the greatest atrocity, horrors, and crimes, such as never has been seen before in the history of all times and of all people."[107] Italian diplomatic efforts in the final weeks had been largely directed at gaining the return of Italian citizens who had already been sent on transports to Poland. Needless to say, nothing came of them. Records from Auschwitz-Birkenau indicate that of the 48,974 Jews who arrived from northern Greece in spring and summer 1943, 38,386 were immediately gassed.[108] After the war, less than 2,000 returned.[109]

THE BULGARIAN ZONE

After the Axis invasion, the Bulgarians annexed eastern Macedonia and Thrace, a region of about seventeen thousand square kilometers that included the provinces of Drama, Kavála, Serres, and Rodopi. About six hundred thousand people total were living there when the occupation began. A year into the occupation Pella and Florina were added to the Bulgarian zone. The region was one of Greece's most economically productive, the heartland of Greek tobacco production, and also the source of a good deal of its grain. As in other regions, all food would be taken by the Axis troops, and in any case, only limited amounts of food were allowed to be transported from one zone to another.[110]

In a context of comparative nightmares, the Bulgarian zone was a particularly bad place to experience the occupation. This was the case for Greek Christians and Jews alike. Bulgaria's long-standing resentment over the loss of Macedonia and Thrace meant that the occupation of those regions was regarded as an opportunity to purge them of all elements in the population that might later compromise the Bulgarian claim that they were rightfully part of Bulgaria. Mois Pessach, from Drama, was among 220 Jews to survive, out of a thousand in the town. He recalls that Jewish sentiment in the region was in keeping with that of the Greek Orthodox. "The Bulgarians . . . were sure that they were going to stay in the region for ever. Like loyal Greek citizens, we [Jews] rejected this categorically."[111]

By summer 1943, tens of thousands of Greek civilians in Macedonia and Thrace had been killed, and a hundred thousand more forcibly removed from their homes. In July, the *Spectator* wrote that the result was that "for the moment Bulgaria's fantastic dreams of holding all Macedonia are realized, except that the Germans still remain in the town and airfield of Salonika."[112] While Jews in the German-occupied zone fled to the Italian zone, Greek Christians in the Bulgarian zone fled to the German zone. "Many have gone to the towns and given themselves up to the Germans. If they are allowed to stay, life under German occupation is much less hard than under Bulgarian."[113]

For Greek Jews, there was not much worse than to be in the German-occupied zone. But the Bulgarian zone came close. In early December 1943, the London *Jewish Chronicle* wrote, "The Germans can truly boast of the thoroughness and astounding rapidity with which they carried out their plan to exterminate the Jews of Macedonia and Thrace."[114] As early as 1940, some of Thrace's Jews realized that Bulgaria's alliance with Germany, along with its proximity to Thrace, was bound to bring trouble. Few, however, made plans to leave before it was difficult to do so.[115] In Thrace, along with the parts of Macedonia that were in the Bulgarian zone, the Germans were greatly abetted by the Bulgarians, who readily turned over all the region's Jews. While the Bulgarians did not turn over Jews living in Bulgaria, in occupied Greece they were happy to comply with German wishes.[116] Some members of the Bulgarian Parliament protested this collaboration, but were reprimanded severely for signing a petition to that effect.[117] A report from Cairo in October had described "the attitude of the Bulgarian authorities towards the Jews, and above all towards those Jews of the Greek provinces that are provisionally being controlled by the Bulgarians" as being "as inhuman and ferocious as that of the Nazi authorities themselves."[118]

As in the German zone, Jews in the Bulgarian one were forced into labor details in summer 1942. Valuables were confiscated and turned over to the Bulgarian government. The Bulgarians were preoccupied with the notion that the Jews were wealthy and beat many of them to the point of death in the effort to learn where they had hidden their supposed riches.[119] In Kavála, in eastern Macedonia on the northern shores of the Aegean, there was a Jewish community of 2,100. It enjoyed close ties with the Orthodox community. Once the town came under Bulgarian occupation, Bulgarian was imposed as the language of instruction in schools. "The Jews registered with the authorities. Their radios were confiscated. Their jewelry and other valuables were deposited in boxes, carrying the name of the owner, at the Narodnaia Banka."[120] Jewish houses and stores were marked. When, in September 1942, the German foreign minister asked Bulgaria to deport all Jews, the Bulgarians expressed opposition. In a perverse sort of bargain, the Bulgarians indicated that they would be willing to arrange for the "relocation" of all Jews in the Bulgarian-occupied zone of Greece instead.[121]

Relations between Jews and Christians in Thrace and eastern Macedonia were better than in many regions of Greece. Gioa Koen, for instance, who spent part of her childhood in Drama, recounts that it was a happy

Fig. 7.5. Snapshot taken by an unidentified person, showing the transfer of Jews of Didimoticho to Alexandropoulis in 1943. From there they were deported to Treblinka and Auschwitz by the Nazis and the Bulgarians. Permission Photographic Archive, Jewish Museum of Greece.

place to live as a Jew.[122] The rapidity with which the Bulgarians turned over the Jews of the region, however, gave them little time to seek help from Christians or escape. While some managed to elude capture—either leaving the country for Turkey, going into hiding with Christian families, or joining the resistance—most were caught. From Komotini, Drama, Kavála, Serres, Alexandropoulis, and Xanthi together, 4,706 Jews were deported—almost the entire Jewish population of those towns.[123]

In Didimoticho, Orestiada, and Soufli in Thrace, all of the Jews were rounded up on March 3–4, marched to the railway station in Alexandropoulis, and loaded, eighty at a time, into boxcars. Simultaneously, arrests were made in Kavála, Serres, Drama, Alexandropoulis, Komotini, and Xanthi. In Kavála, a Greek Christian woman who lived next to a Jewish family, the Simontovs, recalls that the guards who came to take her neighbors away told the Simontovs not to cry, saying, "You won't be separated, you're all going together! We're taking you all on a journey!" Within hours, all the Jews had been marched toward the harbor, among them those who were so old that they could neither see nor hear.[124] In Drama, the whole community was rounded up and sent to the tobacco warehouse, a huge, cold building where they were kept for two days. A few escaped to the mountains; the rest were put in animal carts and shipped through Bulgaria.[125] In Xanthi, home to about six hundred Jews, only about forty survived, some simply through sheer luck; Yuda Perahia, for example, had gone to Komotini on business and was spending the night in Kavála when the roundups took place. Since he wasn't registered with the Kavála community, he escaped deportation. The roundups were so swift and unexpected that only such random happenstances made last-minute escape possible. The members of the Xanthi community were rounded up so quickly, and in such bitter cold, that "they didn't really know what was going on." The Xanthiotes were put on trucks and driven westward to Drama, where they met a train heading north through Bulgaria.[126]

In Didimoticho, the guards beat the Jews, took their possessions, and "shaved the heads of the men and women and removed their clothing, making the men dress up in women's clothing in order to humiliate them." In violation of German orders, the metropolitan bishop of Didimoticho brought the Jews water, while "all the Greeks shut themselves up in their houses, unable to face the tragic scenes taking place before their very eyes."[127] A survivor, Yaakov Jabari from Didimoticho, recounts that "the Gentiles came to say good-bye with great kindness." Some members of the community, like Nisim Alkalai, a teacher at the Didimoticho Jewish School, managed to escape. Escapees from Thrace sought refuge mainly in Turkey. Alkalai, a veteran of the Albanian front, fled with his cousin to the town of Lavara, held by the partisans. Andartes took them to the Turkish border.[128]

Many of the Jews of Thrace were put on a transport to the Baron Hirsch ghetto in Salonika, then on to Birkenau. The majority was killed on arrival.[129] Those in the Bulgarian zone who were deported via Kavála went first to Bulgaria, then by boat up the Danube to Vienna, and then were sent on to Treblinka. There, all but two hundred were exterminated. In late 1943, the Greek government in exile in Cairo, which had contacts throughout Greece, disclosed to the Western press that western Thrace and eastern Macedonia were, in the antiseptic term of the Nazis, completely *judenrein*, Jew free.[130]

ATHENS, APRIL 1941–MARCH 1944

The Jews who fled Salonika for the Italian zone weren't given much time to enjoy their hard-gained relief. When Italy signed the armistice with

Fig. 7.6. Occupied Athens, 1941, with the Nazi flag flying over the Acropolis. Permission Photographic Archive, Jewish Museum of Greece.

Germany in September 1943, the Germans took over almost the entire country and began administering it directly. Eichmann immediately prepared for more deportations, now sending Wisliceny, only just returned from his posting to Salonika, off to Athens to oversee the liquidation of all Jews in the capital.[131]

But Greek-speaking Jews, at least those in Athens, would prove more difficult for the Germans to get their hands on. Athenian Christians were also more likely to regard Jews as fellow Greeks than were Salonikan Christians; indeed, when the deportations had begun from Salonika, the Council of the University of Athens had petitioned occupation authorities that Jews be treated as all other Greek subjects.[132] Athenian Jews also regarded themselves as such. They had more opportunities to hide with Christian families than did Salonikans, and were also more difficult for the German authorities to identify.[133] In Athens, unlike Salonika, Jews did not live in a specifically Jewish quarter, did not have the distinctive Ladino accent that many Salonikans had, and were highly integrated into Greek Orthodox society and fully identified with Greece. A letter, found at Birkenau in 1980 and written near the end of the war, expressed an inmate's final wishes for the disposition of his property and the consolation he found in the news that Greece had been liberated.

In Birkenau . . . I am not sorry that I shall die, but because I shall not be able to avenge myself. . . . If maybe by chance you receive a single letter from our relatives abroad, I ask you please to write them immediately that the Natzari family is extinguished, murdered by the civilized Germans, that's what *Néa Evrópi* calls them, strange, huh? . . . Mitso, take the piano that belonged to my [sister] Nelli, take it to the Sionidou family and give it to Ilia. Now it will be with him, to remind her how much he loved her, and also she him. Every day we wonder if god exists, and even after all this I believe he exists, that god wants his will to be done. I am dying happy, because I know that at this moment our Greece has been liberated. I shall not live, but may they survive[!] . . . My last words shall be, long live Greece. Marsel Natzari.[134]

Another farewell note, most likely written by a Jew from the mainland, possibly Preveza, was found beside crematorium III at Birkenau in Auschwitz, also in 1980.[135] "To my dear ones," the anonymous missive is addressed. "Dimitrios Athan. Stefanidis, Ilias Koen, Georgios Gounaris, and my beloved buddies. Smaru Eframidou of Athens, and so many others whom I always remember. And also . . . to my beloved fatherland "GREECE."[136]

Greece was the site of the last happiness the camp inmates had seen, and they clung fiercely to its memory. Shaul Chazan, a member of the infamous Sonderkommando, which staffed the crematoriums at Birkenau, had a close group of Greek friends, almost all of whom he'd met in Auschwitz (among them was Marsel Natzari, whose hidden last testament was found more than three decades later buried in Birkenau). Many decades later, Chazan was asked in an interview, "What did you talk about [in the camp]?" He replied, "About Greece. Our motherland from before the war. In fact, there were no other topics. You could not think about the future, we were all waiting to die."[137]

The close friendships that Greek-speaking Jews, particularly those of the younger generation, enjoyed with Greek Christians gave them somewhat better chances at finding refuge, escaping, or blending in with Christian society.[138] The Jews in Athens also had the benefit of over a year of reports coming from the north and were under fewer illusions as to what their fate would be should the Germans take action in their city. Athenian Greeks, unlike their coreligionists in Salonika, also had relatively little experience with anti-Semitism. The Athenian Errikos Sevillias writes, "The Jews in Greece did not know what anti-Semitism was, for we had never encountered it in this blessed land."[139] While this made the imposition of the Nuremberg laws all the more shocking to them, it also greatly increased the likelihood of their finding Christian assistance as the Germans closed in. The heterogeneity of the Athenian Jewish community, which was a mixture of Ashkenazim, Romaniotes, and Sephardim, had also encouraged their integration.[140] In the absence of a shared and unified religious cultural heritage, their common language and culture were Greek. Knowing good Greek, however, was not an instant ticket to integration. While by the 1940s most younger Salonikans spoke it well, they were distinguished by their accent, which was readily identifiable due to its heavy Spanish inflection. Many Salonikans who fled to Athens felt vulnerable, knowing that their voices clearly set them apart as Jewish.[141]

From 1941 to 1943, Athens suffered under the same deprivations as Salonika. In the capital, as throughout the country, scores of people were shot by the Germans in reprisal killings, as suspected Communists, or for any one of a host of small and inscrutable reasons.[142] People keeled over in the street from hunger, and thousands died of starvation. The *New York Times* reported that

> an average of 500 residents of Athens and Piraeus are dying daily; the principal streets of the Greek capital are littered with dead and dying, and with only handkerchiefs spread over the faces of the dead by passers-by differentiating the dead from the dying; the crowds of beggars are growing in numbers daily; virtually all children are developing the wasted, bent, rickety legs and swollen bellies of incipient starvation.[143]

A thriving black market, encouraged by the Germans because of the huge profits it brought, meant that food was available only to the rich or the corrupt. In summer 1942, a loaf of bread fetched 1,500 drachmas and a gallon of olive oil cost 19,000 drachmas. The official rate of exchange was 150 drachmas to the dollar, but on the black market a dollar fetched 2,000 drachmas. Reporters returning from Greece wrote that "many people sell their homes and their gold jewelry for food. As many as two thousand real estate transactions a day take place in Athens alone. Most of this loot finds its way into the hands of the plundering Nazis." The death rate had increased by a factor of fifteen within the first year of occupation, and the birthrate dropped to one-seventh its prewar level. No medical supplies whatsoever were available, and veteran amputees made impromptu wheelchairs out of fruit crates. Rather than surrender the ration cards of deceased relatives, families hid the corpses of their dead in gutters or buried them in secret.[144]

As anti-Jewish regulations were imposed in the German and Bulgarian zones in the north, news of the clampdown made its way to Athens via the EAM, the Italian Consulate in Salonika, and Jews who fled to the Italian zone. The Jewish and Greek authorities knew what was coming, and were better prepared to resist it. Already in March 1943, as the transports began leaving Salonika, various Greek Orthodox leaders began taking steps to ensure that Athens' Jews would not be affected. Four leading Athenian businesspeople wrote to the minister of finance to protest the deportations and lament the grave damage they would do to the Greek economy. That same week, Archbishop Damaskinos of Athens—the only head of a European church to officially condemn the German occupation's treatment of Jews—and twenty-eight other prominent community leaders sent a letter to Konstantinos Logothetopoulos, the quisling prime minister of Greece, to protest developments in Salonika. The next day, they sent a similar letter to Gunther von Altenburg, the Reich's plenipotentiary in Greece.[145] Logothetopoulos did nothing to help. At the first postwar World Zionist Conference, held in London, the Greek delegate, Robert Raphael, reported both on Koretz's "cowardly" betrayal of the Salonikan Jews and Logothetopoulos's collaboration with the Nazis. When "all the intellectuals of Athens . . . asked Logothetopoulos to do something [to intervene] with his friends the Germans," and when the president of the Jewish Community Central Board also approached him "with tears in his eyes, [and] asked him not to send the Jews to the death camps," Raphael recounted, "[Logothetopoulos] did nothing."[146]

For several months the situation was uneasy, but not untenable. A report from Athens from the end of September indicated that thus far, the greatest burden placed on the city's Jews was a "crushing compulsory taxation" of Jewish merchants.[147] This was on top of the fact that the tax burden had already been increased for all citizens, who had to pay exorbitant donations and inheritance taxes, profits taxes, and a special tax on agricultural production in addition to the reintroduced tithe.[148] At around the same time, the Italian zone ceased to exist, and the situation rapidly worsened.

Just weeks after the German takeover of the Italian zone, all Jews in the city were given five days to register with the German authorities. Whereas in Salonika compliance with this regulation had been nearly unanimous, in Athens less than one-seventh of the city's eight thousand Jews ultimately complied.[149] In the initial five-day period only about sixty did so. This utter failure, from the German standpoint, stemmed largely from the fact that the Athenians had learned from the hard experience of the Salonikans as well as from the fact that while Salonika's Jews had been easy to identify, Athens' were not. It was also thought that general Athenian opposition to the measures made the Germans "proceed less vigorously" than they had in Salonika.

Athens' Jews also had more possible routes of escape available to them. The EAM was a powerful advocate, and arranged for many Jews to be taken out of the city by boat and truck. In at least one case it devoted an entire ship to the cause of evacuating Jews. The island of Evia was a clearinghouse for escapees who fled the country by sea, mainly to Turkey and Palestine. Some fled on their own to join the andartes in "Free Greece," in the Grammos Mountains in the far north of the country.[150] But life there was not much easier. Elia Aeilon, along with several members of his family, fled north with a resistance liaison who met them at a designated rendezvous point in Athens in fall 1943. The group of nine was passed from guerrilla to guerrilla as they made their way further and further from Athens. In the mountains there was little to eat, and their guides confiscated animals from shepherds and stole grapes from vineyards that they passed along the way. After four weeks in hiding, cold and hungry, living outside in the hills, the group returned to Athens, on the back of a tobacco truck.[151]

Multiple factors—a high degree of assimilation, resistance on the part of Orthodox Christians, a powerful and relatively well-organized resistance movement in the area, and possibilities of escape—made the Athenian Jews unwilling to comply with the German demand that they register.[152] News had also started reaching Athens about the final fate of the Salonikan deportees; reports of "poison gas" in "special bathhouses" were in circulation.[153] The most powerful factor, however, may have been the utterly remarkable and curious disappearance at the end of September of the Athenian chief rabbi, Elias Barzilai.

On his arrival in Athens at the end of September, Wisliceny had summoned Barzilai, telling him that he had three days to produce a list of all Jews in the city. While the Germans had some information on the city's Jews, both from Jewish agents sent from Salonika and German agents posing as Jews, they did not have any systematic detailing of their names or occupations.[154] They knew where Jews lived largely by sending agents after suspected Jews in the street, shadowing them home, and taking note of their addresses, but they didn't have access to any comprehensive lists.[155] When summoned, the British Jewish press recounted that

> in order to gain time, Dr. Barzilai pleaded that the lists were incomplete and obtained a respite of three days for their delivery; but instead of preparing the lists he destroyed all records of the community and "disappeared." Thanks to the respite, Athenian patriots succeeded in causing the entire Jewish population to vanish. Jewish families were dispersed in non-Jewish homes. . . . The Jews remained hidden while the patriots, with the aid of civil and religious officials, as well as of the Greek police, prepared false identification cards enabling the Jews to pass as Christians.[156]

Barzilai had served as president of the community since 1941, and had many contacts in the Salonikan Jewish community and in Athens.[157] When

Fig. 7.7. Louis Koen, ELAS partisan known under the nom de guerre "Kapetan Kronos," photographed in October 1944, Elefsina. Permission Photographic Archive, Jewish Museum of Greece.

he received word on September 21 that Wisliceny wanted to see him, Barzilai immediately notified the city's Jewish and Christian leadership that what had happened so recently in Salonika had now begun in Athens.[158] He also knew that agents had been circulating among the Athens Jewish community for months, gathering what information they could for the authorities.[159] When he did not return with the demanded lists, Wisliceny sent German soldiers to his home. Barzilai and his family had been smuggled out of the city by the EAM, and the community records had been burned.[160] By the time the community's members were required to register, one week later, many had already fled, reasoning that if their leader had seen fit to flee, they should too. But others refused to do so.

Two weeks after Barzilai's mysterious departure, the EAM underground paper *Eléftheri Elláda* published an open message from the rabbi, calling for all Greeks to follow the example of the EAM and extend assistance to Greek Jews in their efforts to elude the Nazis.[161] As the *Jewish Chronicle* reported, "This 'breathing space' enabled the patriots to make arrangements for the rescue of the Jews of Athens. Within a week the entire Jewish population of the city vanished." While overly optimistic—by no means had the "entire" Jewish population disappeared—the report accurately captures the rapidity with which many, many Athenian Jews went into hiding or escaped in the wake of Barzilai's stunning disappearance. More than three thousand escaped with the partisans, hiding out with them in the mountains until the end of the war.[162]

In a later testimonial, in 1954, Barzilai recounted the formal deal that he had negotiated with the partisans on behalf of the Athenian Jewish community. The partisans would feed and clothe the Jewish fugitives, take them to safety in the north, and protect them as best they could. The Jews would not be required to fight with the resistance troops, but if they wished to do so they would be given the same status and respect as all other andartes. In return, Barzilai promised to draw up a certificate testifying to the efforts of the resistance on the Jews' behalf, which the partisans could use in their fund-raising efforts abroad. Jewish assets were to be turned over to the partisans to support the resistance.

When Barzilai was summoned to the Gestapo headquarters in Athens, he knew immediately that it was time to put the agreement into effect. "When I walked out of [Wisliceny's headquarters], I decided to finish the job of rescue. I immediately contacted the heads of the partisans in Athens, and I asked them to arrange the details of the rescue of all the Athenian Jews."[163]

The Greek authorities also intensified their efforts on behalf of the city's Jews. Archbishop Damaskinos famously had clergy announce to their parishes that Christians had an obligation to do all they could to help and issued baptismal certificates to Jews so that they could pose as Christians.[164] The Athens community's history of friendly interactions with Orthodox Christians and the comparatively liberal nature of the Greek rabbinic establishment led many Athenian Jews to take advantage of such measures. While in Lithuania and elsewhere rabbis told congregants that under no circumstances—even in cases of life or death—could a Jew save themself with a baptismal certificate, in Greece the rabbinic authorities not only allowed but encouraged the practice.[165] Angelos Evert, the chief of police, issued an order that false identification cards were to be distributed. Thousands of cards were made up for those who needed them;

"there was not a single Jew, in all of Attica, who didn't have in his pocket a false *taftótita* [ID card]."[166]

On the other hand, the quisling government collaborated with the Germans. Prime Minister Ioannis Rallis organized a special police force of four thousand Greeks who were sent north to look for Jews and their Greek leftist protectors. In Athens, there were also some Jewish collaborators who searched for Jews in hiding and turned them over to the German authorities.[167] For the most part, the property of Jews who went into hiding was left undisturbed, but there were reports of some lootings, mainly of Jewish furniture. It was clear that the situation was fragile. If Greek Christians, who like all Athenians were suffering acutely from hunger, were offered too powerful an incentive by the Germans, it was feared that they might begin to turn in their Jewish neighbors.[168] By summer 1944, more than one in seven Greeks were homeless; over one million displaced citizens searched for shelter, many of them coming to Athens.[169] The stresses of daily life were intolerable; over the course of spring 1944, Athenian Jews increasingly feared that if the situation got much worse, their hitherto largely kindhearted Christian fellow citizens might turn against them. But for the most part, they did not. Reports of what was going on in Athens and throughout Greece came from refugees arriving in Syria, Lebanon, and Palestine.[170] The Palestinian Jewish press followed the events in Greece closely, publishing numerous articles on the aid given to Greek Jews by their Christian fellow citizens.[171]

German efforts also intensified. Jews were required to register with the German authorities and report to them every other day. They were not allowed to change their address, and were to stay inside from 4:00 p.m. until 7:00 a.m.[172] After the flurry of regulations, for several months the situation remained more or less as it was—about two thousand Jews remained publicly in Athens. Some reported for the regular roll call, and others did not. Thus, "the Jews of Athens . . . became separated into two groups, those who were registered and those who were not. The former believed that they were law-abiding and their only inconvenience was that they had to present themselves for a few minutes at the [Etz Hayyim] synagogue."[173]

In marked contrast to the ruthless efficiency that characterized his liquidation of Salonika's Jews, in Athens Wisliceny took no further definitive actions. The registered Jews, however, had to report to him constantly, coming to the synagogue, where they encountered the "monstrous hippopotamus" with his flapping jowls and drooping flesh. Wisliceny's obsessive hobby was photography, and he would force those who came to sign in to pose for him, capturing "the faces wracked with worry . . . [and] anguish" with his Kodak camera. For several months, the Jews duly reported for the regular registrations. Beyond that, nothing further happened. Some grew optimistic, even going so far as to express relief that they hadn't chosen to flee—after all, hiding out in the mountains was not an easy way to live.

In January 1944, Wisliceny was transferred to a post in Theresienstadt. His successor, Toni Burger, was more proactive.[174] Just before Greek Independence Day, on Friday, March 24, 1944, he summoned all the city's adult male Jews to the Melidoni Street synagogue, where armed guards blocked the doors, only letting people out to go and fetch their families. Passover was near, and in an effort to lure as many Jews as possible, the Germans had posted a sign promising free matzo, oil, and sugar to all who came.[175] The announcement was shouted out to the crowd of men gathering in the street.

> All of a sudden someone shouted that we should all get into the synagogue because flour was going to be distributed for the Jewish Easter [Passover]. Since the crowd was large some people remained outside, but a German told them that everyone had to get inside and that is how we got caught like mice in a trap. First of all there was no sign of flour and then, suddenly, the big door closed and two Germans who had been hidden stood in front of it, armed to the teeth.[176]

Some of the men managed to call out to others, warning them to stay away, but few people who had come to register managed to escape.[177]

The three hundred or so men who had first come for flour and registration had grown to a crowd of close to a thousand by the end of the day. Mainly Greek but also Spanish and Portuguese, as many Jews as possible were rounded up in a citywide sweep. They were sent to the infamous Haidari concentration camp in Athens, where they joined about four hundred more Jews imprisoned there earlier. Athens was also the gathering point for six hundred Jews from Preveza, Arta, Agrinion, and Patras. Days later, all the Jews gathered in Athens were put on transports from the Rouf railway station, bound for Auschwitz.[178] At Rouf, "they were packed with suffocating tightness, 80 or 100 at a time, into 37 closed box-cars to be sent north. . . . People were piled on top of one another: pregnant women, children, invalids and old people, all calling for help."[179] The convoy stopped en route in Larissa, where it was joined by 2,400 more deportees from Volos, Trikala, and Jannina. Another 900, Jews from Kastoria, were put on board in Salonika. Altogether, 5,200 deportees in eighty cattle cars traveled through Macedonia, Yugoslavia, Hungary, and Austria. In Vienna, some cars were detached and their passengers marched on foot to Bergen-Belsen. The vast majority went on to Auschwitz, "the vestibule of death."[180] There, they were greeted personally by the infamous "physician" Josef Mengele, who took 648 of them for "research." The rest were put on brutal work details or sent to the gas chambers.[181]

Another group was sent from Athens to Poland in May, following another roundup that included Jews of other nationalities, largely Spanish, Italian, and Turkish. Despite the protests of various ambassadors and the attempted intervention of the Greek Red Cross, they were taken away in "appalling" conditions. "The trucks into which men, women and children, healthy and sick alike, were crammed were normally used for coal and wood. In these trucks, entirely shut off from the outside world, the Jews had to remain for two whole days at the railway station before the train started on its journey." While waiting, some died; one gave birth. Without food and water for forty-eight hours, the deportees were forbidden communication with anyone outside. "The only concession made was to remove temporarily the barbed wire covering a small window in each truck, through which [Red Cross nurses] pushed whatever food they could."[182]

On its arrival in Auschwitz, this last Athens transport was caught up in the midst of a huge series of transports arriving from Hungary. All transports arriving at the time were diverted directly to Birkenau, and the people on them sent straight to the gas chambers, with no selection.[183]

JANNINA AND OTHER MAINLAND REGIONS: THE PELOPONNESE, THESSALY, AND EPIRUS

The deportations of Jews throughout the Greek mainland were timed to coincide with those in Athens. In the Peloponnese, in Preveza, Arta, Agrinion, and Patras, the roundup took place the same night it had in Athens. Jews from Peloponnesian towns were sent on to Athens for deportation. Simultaneously, those from Volos, Larissa, Trikala, and Jannina were taken to Larissa in trucks, where they waited for the trains from Athens to pass by. After the many months of delay in Athens, and the consequent rising hopes of the Jews who lived there, the destruction of Greek Jewish mainland communities was ultimately effected quickly and with a well-timed precision that took many by surprise.

The rate of survival in the towns of the mainland varied considerably according to local factors. Michael Matsas convincingly argues that most critical among them were the actions of the local Jewish leadership, on the one hand, and the relative power of resistance organizations, on the other.[184] The degree of assimilation, while certainly important in the case of Athens, does not appear to have been a definitive factor elsewhere. In Jannina, one of the oldest Greek-speaking Jewish communities in the country, the devastation was almost complete, largely due to the compliance of the Jewish leadership with German demands. In Patras in the Peloponnese, and just to the east in Volos, Larissa, and Trikala, there were coordinated efforts made on the part of the Greek resistance, assistance from Greek Christian authorities, and a relatively low level of compliance with German orders. In these towns, the majority of the Jewish residents escaped.

Volos, Larissa, and Trikala together were home to about 1,200 Jews. In the case of all three towns, the vast majority of Jewish citizens fled, mainly taking refuge in the mountains, before the end March 1944. In Larissa, in fall 1943, when Jews were first required to register, less than two hundred of the town's 1,100 Jewish residents were still there. Rabbi Kassouto impeded German efforts to account for the rest, claiming that he kept no community lists and had no archives listing the names of kehila members. When the Nazis rounded up all of Larissa's Jews some six months later, they found only 225.[185] The proximity of Karditsa, a leftist resistance stronghold, was a bulwark of protection for fleeing Jews from the region. In Trikala, which was dominated by the resistance, only 6 families—out of just over 500 persons—were found and taken to Larissa to join the transport en route from Athens to Auschwitz. In Volos, the vast majority of Jews were saved, a result of the combined efforts of its chief rabbi, Moshe Pessah, and the town's mayor, who colluded together to delay reporting back to the German authorities when they demanded the names of all the town's Jews. The delay facilitated the escape of more than 700 of the town's 822 Jewish residents. To the west, in the Peloponnesian town of Patras, only 12 families were found and sent to Haidari; as in Larissa, the vast preponderance of the town's 240 Jews had fled six months before the roundups of late March 1944.[186]

The news from the mainland was not, however, by any means all good. These "happy" statistics appear as such only when placed against a backdrop of immense misery and incomprehensible cruelty. While the rates of survival were high for many towns, this mitigates none of the horrific suffering endured by those who did not get away. The overland transports that took them to meet trains in Larissa and Athens were conducted under excruciating circumstances. The deportees were subject to unspeakable treatment at the hands of their captors, and having heard ample reports of the previous summer's events in Salonika, they would have had a good idea as to where they were being taken and probably some idea of what would happen to them once they got there.

Perhaps no case better demonstrates the complicated factors that determined the rates of survival than Jannina. In 1944, the city was home to about two thousand Jews, and comprised one of the oldest Jewish communities in Europe. Its existence in Jannina was characterized by a peculiar combination of the radically divergent circumstances that prevailed in other Greek Jewish communities. Like the Athenian Jews, the Janniotes spoke Greek and for the most part enjoyed good relations with their Christian fellow citizens. Like the Salonikans, they were a mainstay of the local economy and lived together in a separate Jewish district of town. But while Athenian Jews had been helped by the fact that they spoke Greek and their ties with the Christian community, the Janniotes were given a false sense of security by their fully Greek acculturation. They also were victims of weak rabbinic leadership, which urged them to obey the authorities and argued against flight during the occupation. This last, and most key, factor is in dramatic contrast to the actions of Rabbi Barzilai in Athens and other Jewish leaders who urged flight and noncompliance.

When the demand that all Jews register came at the end of October 1943, the Greek authorities and the community's rabbi, Shabtai Kabilli, urged them to comply. Greek police officers helped the Germans compile lists with the addresses of Jewish homes.[187] Nevertheless, in Jannina, as in other Romaniote and Greek-speaking settings, it was still difficult for the authorities to know who was and was not Jewish.

> The Germans didn't know on their own who we were, and how many of us there were. . . . [So] the Germans took over the Jewish old folks' home. And they made an announcement that all Jannina residents from twelve and up had to come to the offices at the old folks' home, to get ID cards. As soon as you came in, on the right, in the Community Office, there were two police officers.

As each person walked by, if they were Jewish, the police officer would be told, "Yuda," that's a Jew.[188]

The Germans, having learned from events in Athens that it was crucial to keep panic to a minimum, announced that the Janniotes would not be subject to the Nuremburg laws. It is reported that the SS commander in the city told Jannina's Jews that because they spoke Greek, they would be treated differently—better—than other Jews in Greece.[189] Versions of this story appear in many accounts; it may derive from reports of what Rallis ostensibly told Athens' Jews around the same time: "The Jews of Salonika were the avant-garde of the powers of destruction, and that is why they were destroyed. Not so the Jews of Old Greece. We all know that they are loyal and have a conservative spirit, and therefore they are considered to be real Greek citizens. Therefore the Germans will never harass them."[190] What is clear, though, is that in Jannina, Greek Jews felt themselves to be "more Greek" than other Jews in Greece, and that their superior integration would protect them.

German authorities picked up on the subtle gradations of Greekness that cut across the various Greek Jewish communities, and particularly on the distinction between the more seemingly foreign, Ladino-speaking Jews of Salonika and the Greek-speaking Janniotes. The strong sense of indigenous Greekness that characterized Jannina's community was manipulated against them by the occupation authorities, and later, held against them by other Greek Jews as "naive" and "hubristic."[191]

Reports of escapees confirm that the situation in Jannina and other parts of Old Greece was indeed better than in the north following the arrival of the Germans, both in terms of the restrictions placed on them and their relations with Christians. When a number of Jewish refugees arrived in Turkey, they were interviewed by the U.S. consul Burton Berry. In a report based on the interview, Berry wrote that Jews of Old Greece had respect for Christians, while those from Salonika had felt that the Christians were hostile toward them. "In other parts of Greece, like in Jannina, the Jews do not wear the yellow star, they were not [immediately] required to register, and they have to pay for a permit to leave the city. Otherwise their daily life goes on much as usual."[192] These relatively lax circumstances were attributed to the greater acceptance of Old Greece's Jews as genuinely Greek, in contrast to the foreign, non-Greek Jews of the New Greece.

In any case, for the most part, the members of Jannina's community were reassured; only one hundred Jews left the city in the coming months, half to go into hiding in the mountains, and the others on business. Flight was not easy, either logistically or emotionally. Two brothers who happened to be out of town when Jannina's Jews were forced into custody returned of their own free will and turned themselves in to the German authorities. They couldn't bear the idea that their family might be worried about them. Both died in the camps.[193]

When the roundups took place on the night of March 24–25, 1,832 Janniote Jews were collected in the course of an hour. They were driven in trucks to Larissa, where they boarded the convoy coming from Athens.[194] Despite warnings from Jewish communities elsewhere in Greece, Rabbi Kabilli, later termed "criminal and traitorous" by the Jewish press, had for months told the community to sit tight, comply with orders, and not worry.[195] As a result, few people had fled. The leadership "had hindered the bravest from fleeing to the mountains, and they fell into betrayal . . . they were betrayed like sheep."

On the way to Larissa, the Janniotes started to ask what would happen to them and where they were going. Some suggested that they were being taken away to a work details, that perhaps they would be put into a camp. But "no one thought that they were taking us to burn us. No one, no one said *that*!"[196] By the time they reached Larissa, it was clear enough that nothing good was in store for them. A few Janniotes fled from Larissa

while the Germans kept them waiting for nine hours to be taken on to
Salonika. Empis Svolis and his friend Giako waited for the changing of
the SS guards, and then fled for the mountains. Svolis hid out with the
resistance in the mountainous north until the liberation.[197] Almost all,
though, were stuck. The trains continued to Salonika, where nearly a
thousand more Jews, the last holdouts from that city and Kastoria, were
put on board.[198]

Some Christians tried to help Jannina's Jews—most famously Mayor
Demetrios Vlachides, who cared for the synagogue's possessions during
the occupation and is commemorated in a memorial in the Old Synagogue.
Most, however, were impassive. As elsewhere, the looting of Jewish homes
began almost immediately after the transports had left. Those few Janniote
Jews who returned after the war were not, on the whole, given a warm
welcome, and struggled to regain their property and livelihoods.

Of the Janniotes who survived the trip and then the selection that
awaited them on arrival in Auschwitz, many ended up in Block 27, Lager
D, in Birkenau. There they were housed with Corfiotes, Athenians, and
some Salonikans; the majority of Jews in the block were Greek.[199] Many
did not survive the transport, though. The trip to the camps was terrifying. The conditions on the cattle cars—the crowding, darkness, and lack
of food or sanitation—have been well documented. Psychologically, the
deportees were devastated and terrorized by the fact that they did not
know where they were going, how long it would take to get there, and
what would happen to them when they arrived.

> Where were we going? How long would we be shut in there? Would
> they give us food? My God, how desperate! Some babies never
> stopped crying, sometimes quietly, sometimes screaming. Women
> were lamenting and some of the men took out their prayer books
> and were quietly singing hymns. It was dark in the railroad car, and
> they couldn't see to read, but they kept their eyes open, they turned
> the pages, they knew every line by heart. . . . [P]eople would try to
> peer out through the skylights to find out where we were. As if it
> would have made any difference to know exactly where we were.[200]

Over the course of the twentieth century, Greek Jews had undergone a
gradual, if anomalous, process of nationalization. While other groups
(Asia Minor refugees, for example) became Greek through assimilation
within the physical territory of Greece, Jews became Greek in large part
as a result of their exile from that territory. Only within contexts that
were characterized by mass heterogeneous Jewish populations were they
definitively defined as being first and foremost Greek. First in Auschwitz-
Birkenau and later in the context of postwar Palestine/Israel, Jews of

Fig. 7.8. Emmanuel DeCastro and his youngest daughter, Rochelle, ca. 1940, Jannina, Greece. Rochelle was the only member of the family to survive the war. Permission Kehila Kedosha Janina, New York.

Greek origin were a tiny minority within Jewish contexts dominated by other Jewish groups. While in the Orthodox Christian–dominant context of Greece Jews had variously struggled to establish their status as Greeks, rejected the categorization altogether, or developed complex hybrid identities, in the contexts of Auschwitz-Birkenau and later Palestine/Israel, they were understood by others and later by themselves as uncomplicatedly Greek.

PART FOUR

"The Greeks": Greek Jews
beyond Greece

Chapter 8

AUSCHWITZ-BIRKENAU

> The testimony of the victims is indispensable for any en-
> gagement with the subject of Auschwitz. It is only from
> their perspective that the scale of the crimes becomes
> properly apparent.
> —Sybille Steinbacher, *Auschwitz: A History*

ONCE IN THE CATTLE cars on the way north, even the most optimistic
individuals realized that the outcome of their journey would not be a
happy one.[1] One deportee from Salonika, who was fifteen when his family
was shipped out on April 7, 1943, recounts his father's last hopes fad-
ing away.

> It's difficult for me to identify exactly when my father stopped think-
> ing that they were taking us to central Europe to live under . . . Jewish
> civil democracy. Maybe it was when the soldiers shouted "*Raus!*" at
> us and beat us with clubs to make us get into the wagons? Or when
> they locked us, seventy men and women, old and young, behind the
> heavy sliding door? Or maybe, when we arrived at Auschwitz?[2]

The conglomerate we refer to today as Auschwitz wasn't one camp but
more than three dozen. Birkenau, where the gas chambers and crematoriums were located, comprised four of them, and held at peak capacity
about sixty thousand prisoners, two-thirds of them women. The central
camp, which housed the main administration of the whole complex, held
about twenty-thousand. Others worked at surrounding camps, which
spread out from the center in a large radius, covering an area of more
than twenty square miles. Auschwitz, then, was really a city, in which the
residents of different sectors were kept strictly segregated from one another. Over the four years that it was active, up to two million individuals
were "processed" by Auschwitz—incarcerated or, in the case of up to 75
percent, murdered on arrival.[3] By January 1945, when the Third Reich
began its "death march" evacuation of the camp, only sixty thousand—
perhaps 3 percent—were left alive.

Most, but not all, of the Greek Jews who arrived in Auschwitz-Birkenau in 1943–44 were immediately put to death. From the earlier transports almost all were immediately gassed; later, by 1944, when the

Fig. 8.1. A woman and a girl embark on a deportation train, March 3 or 4, 1943, Thrace, Greece. Permission Film and Photo Archive, Yad Vashem, Jerusalem.

Germans needed the labor, a higher percentage was initially kept alive. Of those not sent directly to the gas, some were hand chosen by Josef Mengele for use in his medical "experiments," and some worked for the Sonderkommando and other kommando assignments. Others lived out their days in the barracks, serving on labor details such as the Aussen-kommando, which spent its days lugging heavy rocks from one place to another, and trying desperately to escape the selections that singled out the ill and infirm for death. Many Salonikans were sent to build the Buna I. G. Farben factories, which sprawled over several acres in Auschwitz III; the first group of Greek Jews was taken to Buna on May 2, 1943, soon after their arrival. Over three thousand Greeks ended up working in the factories.[4]

The Strong "Exotic" Greek of the Camps

First among [the merchants in the camp's black market] come the Greeks, as immobile and silent as sphinxes, squatting on the ground behind their bowls of thick soup, the fruits of their labour, of their

cooperation and of their national solidarity. The Greeks . . . have
made a contribution of the first importance to the physiognomy of
the camp and to the international slang in circulation. Everyone
knows that "*caravana*" is the bowl, and that "*la comedera es buena*"
means that the soup is good; the word that expresses the generic
idea of theft is "*klepsiklepsi*," of obvious Greek origin. These few
survivors from the Jewish colony of Salonika, with their two languages, Spanish and Greek, and their numerous activities, are the
repositories of a concrete, mundane, conscious wisdom, in which the
traditions of all the Mediterranean civilizations blend together. That
this wisdom was transformed in the camp into the systematic and
scientific practice of theft and seizure of possessions and the monopoly of the bargaining market, should not let one forget that their
aversion to gratuitous brutality, their amazing consciousness of the
survival of at least a potential human dignity made of the Greeks the
most coherent national nucleus in *Lager* [camp], and in this respect,
the most civilized.[5]

In the camps, the Greek Jews were regarded as exotic and peculiar outsiders by the Ashkenazim who met them there.[6] They were regarded as
quintessentially Mediterranean, and were teased about their tastes and
habits. One prisoner recalls being rounded up with hundreds of other
Greeks and sent to Birkenau. "In one of the blocks [there] we found a
Blokaltester who spoke Spanish. When he heard that we were from Sa-
loniki he said to us, '*Ijos de putanas! Donde esta la halva y el raki de
Saloniki?*' [Sons of whores! Where are your Salonikan halva and raki?]
We answered, '[Look], we didn't [just] come from Salonika, only from
Buna [in Auschwitz].'"[7]
The "otherness" of the Greeks in the eyes of the Ashkenazim who made
up the majority population in the camps cut two ways. To some, the Greeks
inspired awe with their good looks, ingenuity, and beautifully accented
Hebrew. Some non-Greek prisoners "were impressed with the strength
we'd shown in the war of 40–41, and they considered all us Greeks heroes."[8] But to others, they were inferior, second-class citizens. The divide
between Ashkenazim and Sephardim, in which Sephardim were cast as
inferior outsiders, was to become one of the most striking features of postwar Jewish Palestine/Israel. During the war, it was developing already in
the camps—the first sites of heterogeneous mass Jewish cohabitation.

We, the Jews of Greece, were in a special situation, the reason for
which I never understood. Even though we always filled in for sick
members of the *Sonderkommando*, and we always took upon our-
selves assignments that had been given to others—despite all that,
we were treated with disrespect, particularly [by] prisoners from the

northern countries, such as Poles, Russians, and Czechs. We were always the black sheep, and we always were called insulting names such as "cholera," or "*kurva*" [whore]. One day I spoke with [a Polish prisoner named] Kochak, and asked him to try to put an end to the situation. "I can't help," he said. "They are bastards who view you as people of an inferior level. They are the Ashkenazim, and you are the Sephardim." What a shame! They only spoke to the Greek Jews nicely when they wanted them to give them a bit of gold to trade. . . . I, the Greek, the Sephardic, the "cholera," never refused them that small service.[9]

The Greeks' appearance, their habits, and above all, their inability to speak Yiddish struck the Ashkenazim who dominated the camp as the least Jewish thing they had ever seen. Ka Tzetnik (b. Yeḥiel Finer, 1909–2001), a Polish-born survivor of the Holocaust and one of its most famous eyewitnesses, writes of the shock and wonder that central European Jews felt on encountering the Greeks.[10] In one of his novels, a central European Jew named Hayyim-Idl muses over his barrack mate, a Greek who never spoke.

Professor Rafael . . . doesn't tire of just lying there day and night, all locked up within himself, not saying a word even to his friends. . . . So they are as silent as he. And even if Professor Rafael did speak, he, Hayyim-Idl, wouldn't be able to exchange a word with him. They say the Professor knows ten languages fluently. But he doesn't understand a word of Yiddish. Strange Jew. He was brought here from Greece, where he was born. . . . Not even forty yet, and he speaks all of ten languages. But plain Yiddish, the Mother Tongue—not a word!

The camp's Greek Jews speak in Greek, box for recreation, and look like nothing the Askenazic Hayyim-Idl has ever seen:

Jewish lads from Salonika. Jews. Now who would ever guess that the likes of these are Jews? They had to send him to Auschwitz so he should find out that tawny skins and muscles like that could belong to brothers of his. His own flesh and blood. They can't speak a word of Yiddish, but that "Shalom!" comes out of their mouths like a living verse of Scripture . . . Jewish kids. And what is it they want? Just to live. Greece. Jews everywhere. He had never realized that in Greece the Jews look so Gentile. Jewish kids.[11]

The Greeks, who knew no German or Yiddish, suffered from the language barrier that stood between them and the German and Polish guards. "There were two reasons why the Greek Jews suffered more than the other Jews [in the camps]. First, they were Mediterranean. . . .

Second, they came to Poland without knowing the language. The Ashke-nazic Jews who know Yiddish, understood German. We arrived with absolutely no knowledge of the German language, and it was very diffi-cult for us to get used to the central European climate."[12] When Greeks were slow to react to German orders, they were punished. Many were beaten to death simply because they couldn't make out what it was they were being told to do.[13]

Not surprisingly, Greeks sought one another out so they might have some verbal human contact. Errikos Sevillias, "a simple man born in Athens in the good old days of 1901," had served in the Greek campaign in Asia Minor for three years and then returned to Athens in 1923 to open a leather goods workshop.[14] He was deported to Auschwitz in March 1944, where he was assigned inmate number 182699. In his bunk were "a Russian, a Pole, a Hungarian, and the fourth was Dutch. Try to communicate in a situation like that." When a guard asked him his num-ber and he didn't understand the question, "he slapped me on the face and pointed to my arm." Finally he sought out a new bunk "with two Greeks and two Italians and joined them as the fifth. At least I could talk to them."[15]

The language barrier also ensured that few Greeks tried to escape the camp. "Usually it was the Poles who tried to escape. How could a French, Dutch, or Greek prisoner think of escape without knowing a word of Polish?"[16] At the end, too, when the Germans abandoned the camp in the bitter cold first weeks of 1945, Polish prisoners walked away to find food and shelter, while most of the Greeks did not. "But us Greeks, where could we go?"[17]

In the camps, Ladino was a Greek language—a language spoken by Greeks. In Greece, it had marked the Sephardim as outsiders, but in Auschwitz it identified them definitively as Greek. "One time I saw a group of Greek women screaming; the kapo was beating them. How did I know they were Greek? They all were wearing the same clothes, and they were all bald. And it was very difficult to identify individual faces. But I heard that they were screaming 'Dio!' which in Ladino means 'God.' And then I understood that they were Greek."[18]

But while the Germans couldn't understand Greek, they were interested in the strangeness of the language. As Shaul Chazan recounts, "We slept in a block in Birkenau, number 11 or 13. From there we went every morn-ing on foot to work. When we returned to the block, we were forced to sing. We used to sing songs in Greek, folk songs. The Germans really liked the sound of the Greek language."[19] Leon Koen recalls a young German leading a newly selected Sonderkommando squad, made up entirely of Greeks, off to work. " 'All the Greeks, after me,' . . . When we got out of the block, the German asked, 'You know how to sing, right? Why don't you sing a song?' And then we started singing. There were always songs that we liked to sing together—Greek folk songs or patriotic songs."[20] Shabtai Hannuka, a Salonikan who was subjected to Mengele's experiments while in Auschwitz, later recollected, "We used to sing Greek army songs, military Greek marches. The Germans enjoyed listening to us."[21] Testimony after testimony refers to the German guards' fascination with Greek music.[22] As for the Greeks, it provided them with an ironic mode for expressing the tragedy of the Holocaust. Here the language barrier was a boon. "What helped me in the camp was singing. The Germans always told us, '*singen*,' and we used to sing. We sang all sorts of songs; we made up the words. For example . . . 'With laughter on our lips, we haul a cart of shit, and the German behind us beats us with a belt. La, la, la, la, la.'"[23] One of the most biting and poignant relics of Greek Jewish life was the preservation of Greek popular songs of the 1920s and 1930s, recast with lyrics that described life in Auschwitz. Many were from the genre known as Rembétika, a musical style that arose largely in the poorer immigrant neighborhoods of Salonika and Athens in the 1920s, and is one of the few artifacts of the brief, oftentimes—but not invariably— fraught coexistence of Asia Minor Greeks and Jews:

> I didn't know prison, now I do
> Trapped in the cell, I stare at the walls
> All comes back to my mind, the laughter and the loves
> All became ashes, on the train of life.
>
> That's the way life goes, girls, that's the way life always goes
> For us to be closed up in Auschwitz.
> Youth that passes, joys that leave and don't come back.
> Girls, be patient, we'll get out
> Of Auschwitz.
>
> Τη φυλακή εγώ δεν ήξερα / και τώρα τη γνωρίζω
> Μες στο κελί γυρίζω / τους τοίχους αντικρίζω.
> Όλα στο νου μου έρχονται: / τα γέλια κι οι αγάπες
> Όλα γίνηκαν στάχτες στο τρένο της ζωής
>
> Έτσι είναι η ζωή, κορίτσια, / πάντα έτσι είναι η ζωή
> Νάμεστε κλεισμένες μες στο Αούσβιτς.
> Νιάτα που περνούν, χαρές που φεύγουν / πίσω δεν γυρνούν.
> Κορίτσια, κάντε υπομονή, θα βγούμε
> Από το Αούσβιτς[24]

    The inability to communicate with others furthered the sense of solidarity between the camp's Greek Jews. Eli Wiesel writes, "I remember in the camp, in our block, there were Jews from Thessaloniki. They didn't understand my Yiddish and I didn't understand Greek or their 'Ladino.' But I liked to be around them all the same. They had good hearts. . . . The solidarity between them struck all of us as amazing."[25]

This solidarity characterized relations between Greek Jews from varied locales in Greece, and reached across the long-standing Romaniote-Sephardic divide, which had in any case decreased over the prior thirty years, as Thrace and Macedonia were further integrated into Greece. After Greece was drawn into the war, from 1941 to 1943, the emergence of a collective, distinctly Greek—as opposed to regional—Jewish consciousness was further hastened by the occupation and particularly by the tripartite administration of the country, which forced Jews living in the German- and Bulgarian-occupied zones to depend on contacts they had in the Italian zone. While for the first months almost all Greek Jews in the camp were from Salonika, when transports from other regions began to come in the Salonikans greeted the newcomers as fellow Greeks. The experience of the camp, where all Jews from Greece—Romaniote, Salonikan, and Ashkenazic alike—were referred to by the guards and non-Greek Jews simply as "the Greeks," furthered this collective nationalization.

For the few Greeks who could speak and understand German, the ability was a godsend. Erika Kounio Amariglio, who came to Auschwitz from Salonika on the first transport out of the city, writes that of the twenty-eight hundred people on board, only four—her family—could communicate freely with the Germans. As a result, they were set aside as the rest were put in trucks and taken away. " 'Coincidences' had saved us. . . . What a coincidence it was that of the 2,800 people in the first transport, there was no one else who could speak German. And the Germans needed people to act as interpreters."[26] Iakovos Stroumsa, the famous "Violinist of Auschwitz," also knew German well.[27] Sent to Auschwitz-Birkenau on a transport that left Salonika on May 8, 1943, Stroumsa survived the camp in large part because he spoke near-fluent German and played the violin beautifully. In his memoir, he recounts the surreal experience of auditioning on the violin before the Blokaltester soon after his arrival. "What do you want me to play? Mozart, Beethoven, Haydn? Concertos or sonatas?" After playing for twenty minutes, the guard told Stroumsa he hoped he'd survive, because he played so well and because his German was so good.[28]

The Greeks in the camps didn't stand out only for their language and music. They were also regarded as particularly strong, resourceful, and handsome, and emerged as leaders in the camp. Dario Akounis, a stevedore from Salonika, arrived in Buna with a broken leg. He was afraid that if he went to the doctor, he would be sent to the selection. Finally, though, he did see the doctor, who operated, without anesthesia, on the broken bone. The assistant at the operation was a Spanish political prisoner, and as he held Akounis's legs during the surgery, the two conversed in Spanish. The Spaniard asked Akounis how he knew the language. "I'm from Saloniki," Akounis replied. "Oh, yes," the Spaniard responded, "All Salonikans are strong."²⁹ It was good to have Greek friends in the camp. When Primo Levi left Auschwitz in early 1945, he traveled in the company of a man he referred to simply as "my Greek." Levi chose his companion largely on the basis of the fact that he was Salonikan, "which, as everyone in Auschwitz knew, was equivalent to a guarantee of highly skilled mercantile ability, and of knowing how to get oneself out of any situation."³⁰

The Greeks in the camps had a well-deserved reputation for resourcefulness, a quality that could literally mean the difference between life and death. For example, prisoners carried their thin soup in a bucketlike pot called, in the argot of Auschwitz, a *menaschka*. These pots could only be obtained through complicated negotiations with the tinsmith, who would make them for prisoners in exchange for bread. The Greeks had the largest pots of all, which in addition to securing them more food, also was a symbol of their primacy: "Besides the material advantages [of the *menaschka*], it carries with it a perceptible improvement in our social standing. A [large] *menaschka* . . . is a diploma of nobility, a heraldic emblem."³¹

The guards regarded the Greeks as gifted athletes and often deployed them for entertainment. Yitzhak Koen, who worked on the construction of Lager D in Auschwitz, recalls that one day as the workers were cleaning some pools of water used to breed fish outside the camp, "an SS man came and told me, 'You are Greek, Greece is a beautiful country. You probably know how to swim, climb up a tree, and dive headfirst into a pool. So let's see how you swim.'" Koen was wearing heavy boots, and knew that if he dived headfirst into the water he could kill himself, so he jumped in foot first. As punishment for the disappointing athletic performance, the SS officer sent his dog into the pool to bite Koen.³² Greeks were meant to be more athletic than that, reasoned the guard.

The Greeks became particularly well-known for their boxing abilities. The famed Yaakov Razon, a Salonikan who had been coached by Dino Uziel, the Greek national boxing champion, found the skill to be indispensable in the camps, respected as boxing was by the guards.

> When we arrived in Auschwitz, we immediately came to grasp the situation. I exclaimed, "I'm screwed." One of the kapos hit me, and I immediately hit him back. He was stunned and asked me if I was a boxer. I said yes. The next day they had already organized a match for me with a Polish guy. I knocked him out by the third round. . . . When the Jews saw that I was winning, they threw bread, margarine, and a little bit of cheese toward me, and I understood that this was my chance to survive.

When Razon was moved to Buna, he announced to everyone that he was a boxer. The guards matched him up with different fights and even provided him with an arena. "They loved this game, because it is a cruel game. All the SS men used to come watch us, every Sunday we would box." In Buna, Razon coached other boxers and used the privileges granted him by the guards to cadge extra rations for the other prisoners.[33]

The Germans readily recognized that such physical and psychological resilience also had utility. The Greek reputation for strength likely accounts for the fact that large numbers of them were recruited for the Sonderkommando, the squads of Jews who were assigned the work of disposing of the bodies of the Nazis' victims. Leon Koen recalls that when a French-speaking kapo came looking for strong men, Koen told him that the Greek Jews in the block were "ready for any hard work." The following day, they realized that they had been lured to work in the Sonderkommando, the most fearful work detail at the camp.[34] When the kapo led them off to work, he explained that he had taken them "because you all came on the same transport and because you're all Greek."[35]

The Sonderkommando

Many Greeks were made to work in the Sonderkommando, whose branches staffed the various tasks of the crematoriums.[36] So long as they were working, they managed to prolong their own lives. They were given better clothing, lodging, and food in exchange for the horrific work.[37] But most knew that they, too, would ultimately meet the same fate. Berry Nahmia, a Greek survivor of Auschwitz-Birkenau, recounts hearing, from the crematorium opposite her barracks, a Greek prisoner singing as he worked.

> Greek girls who hear me, tra-la-la-la-la. So that you'll understand, I'll tell it to you in a song. The chimneys that you see here so high are the worst factories of death. Thousands of Jews, old, young, little children, fall into the arms of the fire. I know, they'll burn me too, soon I shall exist no more. I will describe the same thing that my tired eyes see. Do you hear me? Believe me, it is horrible, and true. I live it every day. Greek girls, I beg you, if you get out alive some day, from this Lager, go tell the open world, so it can learn of what I sing.[38]

The longer-term Greeks of the Sonderkommando took it on themselves to let Greek newcomers know what really went on in the camps.[39] When the Ioanniotes had just arrived, they passed by the barracks of some of the Salonikans. A few climbed out the windows and walked over to the new arrivals.

—Girls, do you know Greek?
—We do. Where are our parents now?[40]
—Lean over, lean over a bit further. What do you see?
—What do I see?
They wanted me to see the flame from the crematorium that was
burning. And then they said to me:
—There, there they all are![41]

Eliki Sardas, from Corfu, was also introduced to the ways of the camp
by Salonikans she met when she arrived in Auschwitz. "I was afraid of
selections all the time, after they explained to me what it was. We didn't
know the language, and this was really hard. There were some Salonikans
there, and they warned us in Greek that we shouldn't say if we were sick,
because they would kill us. We listened, and some of us were saved."[42]

Yaakov Gabai, who had come to Auschwitz from Salonika in mid-
April, was chosen for the Sonderkommando the month after his arrival,
on May 12, 1944. After a thorough medical check, he and 750 other
men—those found to be the sturdiest and in the strongest health—were
pulled aside by the guards and told, "From now on you will work hard,
but you are not going to lack for food and clothing." They were happy
to hear the news, but didn't know what it meant. "This way," Gabai
recounts, "we effectively became the Sonderkommando." The team was
cut off from the rest of the camp, and was housed in the attics of crematoriums I and II, and inside crematoriums III and IV themselves. On arriving
in their lodgings, the newcomers learned from previously selected members of the command that "the meaning of the Sonderkommando is to
burn corpses every day. This was the first time we heard that they burned
people in Auschwitz."[43]

They were soon divided into groups; Gabai's was made up mainly of
Greeks, and included Corfiotes, Athenians, and Salonikans. Soon enough,
they learned what their new duties entailed. A transport of more than two
thousand Hungarian Jews pulled into the station. Just minutes later,
Gabai and his coworkers were shocked to find themselves looking at dead
bodies, piled one on top of another, in the "showers." The Hungarians
had been gassed on arrival. Gabai was horrified. Soon, though, he knew
he must protect his sanity and ceased tormenting himself with the thought
of how these people had died. "This was going to be our job and we had
to get used to it. Hard work, but you get used to it." And hard work it
was. Gabai's task was to lift the bodies onto stretchers and then throw
them with pitchforks into the ovens. "Each oven had three doors; through
each door you could put four bodies—sixty bodies in fifteen minutes.
After fifteen minutes, you stirred the fire up with a pitchfork. After another fifteen minutes, there was nothing left of the victims but ashes."[44]

While the bodies were burned, "the fats they produced were drained off through pipes into a large pit behind the furnace."[45] The whole process ran with precision, and the members of the Sonderkommando tried to get lost in the routine and monotony of the work.

But it wasn't always possible to maintain a demeanor of distance. One day, a group of four hundred *"muselmanner"*—zombies, the walking dead who, exhausted and extinguished, had no hope of eluding the selections—were brought to the gas chambers.[46] Among them were two of Gabai's cousins. He sat talking with them for two hours, in the dressing room of the gas chambers. After sharing some food and cigarettes, a guard said, "Now it's time to finish you off." Gabai told his cousins,

> "Come with me." I took them into the gas chamber, to the precise place where the gas entered the chamber. I told them, "If you sit [right] here, you won't suffer for even one second." . . . Among the victims that were killed that day were ten of our friends and relatives from Greece. When we finished burning the 390 [others], we burned each of our relatives and acquaintances alone. Then we collected the ashes of each one of them, put them in boxes, wrote down the name and birth date of the deceased, and the date of his murder. Then we buried the boxes and said kaddish over them. We asked ourselves, "Who is going to say kaddish over us?"[47]

While the language barrier protected most Greek Sonderkommandos from having to talk with those brought to the chambers to die, sometimes they met people who spoke their language. Shaul Chazan, another Salonikan member of the Sonderkommando, encountered a former neighbor of his, brought from Buna to the crematoriums. "He wanted to know exactly how he was going to die. . . . I asked him, 'Why [do you want to know]? What will come of it?' But the man insisted, 'No! You must tell me, so at least we will know.' " Chazan recalled, "The fact that he was going to die didn't bother him. He just wanted to know the details. So I told him, it's like this: you undress down there, in the basement, with the others, and then you all go to the gas. The gas will finish you all, and then the corpses will be burned. He wanted to know, so I told him."[48]

More often, the workers could not bring themselves to tell the condemned what it was that was about to happen to them. One member of the command, a Corfiote named Papo Yosef Baruch, learned one day that a transport from Corfu had brought members of his family to crematorium VI. He asked permission to go see them, and it was granted. When he came back from talking with his relatives, his coworker asked, "Yosef, did you tell them that they are going to be murdered?" He answered, "How could I tell them such a thing? I was simply unable to."[49]

By definition, no one who was sent into the gas chambers came out alive. One Salonikan, however, was an exception, and provides one of the only eyewitness accounts of how the gas houses looked from the inside:

> It looked like a huge warehouse, with a concrete floor, with no windows. The ceiling was full of faucets. I thought I was in the showers. We were there more than three hundred men and women, and each of us was holding a towel and a bar of soap. Boy, were we stupid! Fortunately for me, there I met Jacko Maestro, who saved my life. He worked there as a translator, and he managed to arrange a new life for me. He simply pulled me out of the chambers.[50]

While some staffed the crematoriums, other groups within the Sonderkommando had as their job to pull out the teeth of the freshly gassed corpses. Nicknamed "the dentists," they were sent into the chambers right after the executions. "Before they threw the bodies into the ovens, I had to look into the mouths of the murdered and pull out their gold teeth. The Germans did not want to give up the gold." The group was given special dentists' pliers for the job. The Germans did not want anything of any value whatsoever to go to waste. Leon Koen, a Salonikan who was deported from Athens, was derisively called "the Greek dentist" by the German guards in the camp. Koen recounts that once, when some gold teeth were found among the ashes in the crematoriums, he was lashed with a studded whip and accused of "sabotage." The dentists extracted the teeth and put them in small bags that were attached to their boots. They then called another prisoner, whose job it was to melt down the teeth into gold bars, which he gave to the Germans. "We nicknamed [him] the treasury minister." When they could, some members of the teeth detail hid away gold teeth, trading them on the guard-run black market for special food.[51]

Female Kommandos also were assigned to various labor details. The so-called Kanada Kommando, which included many women from the Greek transports, was responsible for sorting through the clothes and luggage that the deportees had so pointlessly brought with them. Others, such as the Union-Werke Kommando, worked in a munitions factory. The workers there would conspire to have their fellow nationals employed as well, so large groups of Greek women ended up working there together.[52] They also served as secretaries and ran the bureaucracy of the camps.

The women who worked in the Union-Werke Kommando colluded with the Sonderkommando to bring about one of the greatest acts of mutiny to take place in German camps during the war. A number of its leaders were Greeks, among them the Corfiote Papo Yosef Baruch. The women who worked in the munitions factory had slowly, over the course of weeks, been smuggling out explosive material and passing it on to the members of the Sonderkommando, who hid it beside the ovens. On October 7, 1944, as the inmates of Auschwitz-Birkenau grew hopeful that the end was near, one team of the command managed to set fire to or blow up crematorium III, killing some SS in the process. The Sonderkommando uprising in Birkenau was the third such event, following similar events in Treblinka and Sobibor. It was the only organized, armed act of resistance in Auschwitz.

The Sonderkommando, cut off as they were from ready contact with the rest of the camp, nevertheless managed to establish ties with a camp-wide resistance organization, made up of both Jews and non-Jews, that had been established early in the camp's history, in 1942, by Polish prisoners. Together, the groups made plans for a large revolt. The Sonderkommando rebellion was to be part of a campwide uprising; initial expectations were that over fifteen thousand prisoners would participate, and that hundreds of armed Polish resistance fighters just outside the camp would come to the rebels' assistance.[53]

Such hopes proved optimistic. Conflicts arose between the main resistance organization, which wanted to take time to plan a large organized movement, and the Sonderkommando, who wished to move more quickly, partly in an effort to save the lies of the Hungarian Jews who were being shipped to Auschwitz in summer 1944; by mid-July, over 437,000 Hungarian Jews had been gassed in Birkenau.[54] As a result of the conflict, the Sonderkommando rebels started to make plans independently of the planned general uprising. The fact that the Sonderkommando were among the few in the camp who realized that the Nazis did not plan to leave anyone alive lent urgency as well. While others in the camp held out hopes of survival, the Sonderkommando knew that no eyewitnesses would be allowed to live.

Earlier uprisings, in June and August 1944, had been aborted when the general resistance organization deferred them. At the end of June, most of the Sonderkommando were moved from their block in the men's camp into the crematoriums themselves, in order to increase their isolation from the rest of the prisoners. "But the flame of rebellion did not die in the hearts of the prisoners from Greece."[55] In August, the SS killed the chief kapo of the crematoriums, Yaakov Kaminsky, who was one of the organizers of the impending rebellion. On September 23, almost two hundred of the Sonderkommando were themselves murdered by the Germans, who had apparently gotten wind of a planned attack. These events, too, prompted the Sonderkommando to act fast. Days before the uprising, the Sonderkommando underground heard that its remaining members were to be exterminated in the coming days. It seems that this information was accurate. Lists of Sonderkommando members were made up, ostensibly for the purpose of "transfer." When the general resistance told the Sonderkommando resistance not to take action, they ignored the orders. As it became clear that some of the SS appeared to have heard word of the planned rebellion, the Sonderkommando moved ahead.

The exact sequence of events is unclear. The rebelling prisoners used improvised weapons to attack the guards. Three members of the SS were killed, and some were wounded. A handful of rebels managed to get inside crematorium III and set it on fire. Some witnesses differ in their reports, saying that the rebels were in fact able to blow it up. Another squad, seeing the events from afar, by crematorium I, thought a general rebellion had broken out and proceeded to rebel as well. Even though 451 prisoners were killed in the uprising, 212 of the Sonderkommando working at the crematoriums that day survived.[56] The surviving workers were rounded up and were sure they were going to be killed because of what had happened. "The treasury minister managed to smuggle us some bottles of whiskey and some food. Emboldened by the drinking, we started singing Greek nationalist songs and *Romanceros* in Ladino, which were so dear to the Jews of Saloniki."[57] To their astonishment, this group was left alive.

The only attempted escape that resulted in the death of German officers had been made by a Greek, and his bravery became an inspiration to others in the camp, some of whom participated in the Sonderkommando rebellion less than two weeks later. Albertos Erreras, an army veteran from Larissa who during the occupation had taken the name Alekos Michaelides, was assigned to the labor detail that shoveled the ashes from the crematoriums into the river that ran past the camp. The workers on this detail were accompanied by two SS guards, both of whom Erreras managed to kill by smashing them on the head with his shovel and then throwing them into the river. Erreras crossed the river and fled into the woods on the other side, but ultimately was found by German dogs. After prolonged torture, Erreras was executed and his body prominently displayed as a warning to others who might try to escape. The main example he set, however, was one of bravery. The reputation of Greeks in the camp was further enhanced by the episode, as well as the Sonderkommando revolt, after which even the Germans showed some signs of admiration.[58] Greeks, in particular, later remembered the rebellion as a moment of Greek national glory.

The few prisoners from the Sonder who took part in the rebellion and survived later became a legend. The legend was prevalent even among the *Schupo*, who defined us as "Greek bandits," but admired us. . . . A long time after our escape from Birkenau [January 17, 1945] we met prisoners in other camps who asked if what they'd heard about the Birkenau rebellion was true. They told us how much they had been encouraged . . . when they heard that a group of Jews from Greece came out against the Goliath of our time, and that their people knew how to die with weapons in their hands, bringing honor to their people and their country as Jews and as Greeks.[59]

MEDICAL "EXPERIMENTS"

The infamous German physician Josef Mengele was in charge of the selection of Jews arriving in the camps. Many of the arriving Greeks found "the youthful doctor" waiting for them on the platform at Auschwitz when their transports finally came to a halt.[60] Mengele came to Auschwitz in May 1943. He determined who was able-bodied enough to work and who was to be sent directly to the gas. He also was in charge of conducting scientific experiments, most famously on twins and very small persons. Another of his preoccupations was with fertility and reproduction. On their arrival in Auschwitz, a number of inmates were examined by Mengele and his team to see if they were appropriate subjects for his experiments.

The inmates set aside as the subjects for these experiments made up a separate Kommando; one group came to be known among the other prisoners as "the Kommando of Fifty," a group of that number comprised of young boys seventeen and eighteen years old. "One day, after about two weeks, they were taken to a workshop where they were greeted with electric wires . . . put on their testicles: they became sterile! From that day on, 'the Kommando of Fifty' was called 'the Kommando of the Sterile.' "[61]

The Nazis gave most women in the camps medicine to suppress menstruation. It didn't work on everyone, and for the women who did continue to menstruate it was humiliating, given the lack of privacy and sanitation. The medicine made everyone break out in a terrible rash. The pseudomedical tortures to which inmates were subjected were a particular source of anger and suffering; one survivor from Jannina recalled that above all it was for these humiliations that she hated the Nazis. "If I could take their eyes out. If now I had a German, man or woman, here before me and I could slaughter them, I'd do it. If I could take their eyes out."[62]

Twins—particularly "good material"—were subjected to awful and bizarre experiments. In one "study," they were made to lie on an ice-cold floor while a German doctor, "civilized, with a university education," looked at his watch to see how long they could survive the violent shivering of their own tiny bodies. "*Óla yiá tin epistími!*" [Everything for the pursuit of knowledge!] later remarked an eyewitness, in grim irony.[63]

When his transport arrived from Athens, Leon Koen recounts, "Dr. Mengele was waiting for us there. He was a young man in his thirties. Next to him stood a woman . . . and two gigantic German shepherd

[dogs]. He started the selection—left—right—left—left. On one side the young men were concentrated, and the women, children, and the elderly on the other."[64] When he found young boys who were particularly virile, he would have one of their testicles removed to examine the effect. About a month later, they would be summoned for further test taking. Then their other testicle would be removed.[65] Newcomers who were warned in time rushed into the latrines to masturbate, in the hopes that on examination by Mengele's assistants they would be found lacking in potency and thus uninteresting.[66]

Block 10 is where Mengele and his colleagues performed many of their experiments. "Women were housed in one wing, and we heard that the majority were Greek."[67] When the Greek transports were at their peak, in 1943–44, they were scoured by Mengele's team for virgins and newly-weds. "For some reason the Germans chose the Greek Jewish women to experiment on."[68] "They experimented [on them] in hopes of finding ways to sterilize men and women faster and more effectively. This was basically Dr. Mengele's specialty." Generally, no anesthesia was used while the experiments were being conducted.[69] Women were given injections in the uterus, subjected to gynecologic examinations, and sterilized. Most survivors of the block were unable to have children later.[70] Many women spent more than a year in Block 10, passing their days in a blur of fevers, pains, and narcosis.[71]

Other so-called experiments were phrenological. The Germans who conducted them made plaster casts of the heads of the various supposed races interned in the camp. The plaster heads were lined up next to one another for comparison. "At the base of each . . . was a piece of cardboard with the identification 'Gypsy,' 'Ukrainian,' 'Polish,' and others."[72] Greek specimens were in hot demand.

Just as centuries before on Corfu the Greeks had defined themselves against other Jewish groups, in the multicultural setting of the camps Greek Jews affirmed themselves vis-à-vis other Jews as Greek, rather than as Jewish. For most of them, who had lived for centuries in the company of Christians, and had been defined against them as emphatically Jewish and somehow not really Greek, this was a largely new identity. But in Auschwitz their Greekness was assumed, not something they had to fight to assert or deny, as it had been in Greece itself. They had lived in Greece and were deported from Greece—they were Greek. This Greekness became a badge of honor and mark of prestige in the camps. Greeks—in many instances, the first non-Ashkenazic Jews that the mostly Ashkenazic inmates had ever seen—were virile and strong. They were silent and mysterious, ingenious and tricky. They were fiercely patriotic, and came together with a sense of national unity that was alien to the others.

In some cases, the bond felt between Greeks in the camps was so over-powering that it transcended other, more personal ties. In the horrible anonymity brought by the starvation and tattered uniforms, which made everyone unrecognizable and alike, being Greek filled in for other forms of familiarity. Just before the liberation of the camps, Naoum Negrin of Jannina found himself huddled together with a bunch of other inmates in the block.

[A man was brought] into the block where we were so that he could warm up, because it was winter, really heavy winter. The others talked to him, asked what was going on, who he was, but he couldn't answer what they asked, not in German, nor in Hungarian. We were no more than ten people all together. Someone said to me, you ask him. I asked, *Italiano tu?* Because since I'm looking at him but don't recognize him, I figure he won't be Greek. He's been in there for half an hour with us, and at some point they wanted to tell him to leave. He had to leave, because he wasn't from our barracks. When I said, "Italiano tu?" he said, "*Ohi.* [No.] Greco." The others listened to the exchange. "OK," I said. "A Greek from which place?" "From Jannina" . . . I said, "But I'm from Jannina," and I asked him what his name was. "Samuel Kantos." I knew him by the name of Melias, and we had been neighbors. . . . We embraced each other. . . . The others looked at us and asked, what's going on? Bruder, I told them, brother. "You're brothers and you don't know each other?" I sud-denly realized we'd been reduced to skeletons.[73]

In the anonymous, atomized existence of the camps, common Greekness was just about the only thing left that tied the Greek survivors together. Almost all had lost any other semblance of family ties; many were the sole surviving member of previously large families. Understand-ably, they sought out comfort in other Greeks. Esther Rafael, from Corfu, was taken from Auschwitz to Theriesenstadt after the liberation, yet was so dazed that she didn't realize that it was no longer a concentration camp, but had been converted into a displaced persons camp. She walked from infirmary to infirmary, opening the door and calling out "Grecos?" In one, a man, weak with typhus, raised his hand. They spoke Greek together, and while they didn't know anyone in common, they quickly formed a bond. Soon after, the two were married. "Today this man is my husband, Hayyim Rafael."[74]

As the end of the war approached, the Nazis rushed to liquidate huge numbers of Hungarian Jews, among the last to be sent to the crematori-ums. The camp was abuzz with preparations. Yitzhak Koen of Salonika

was on the labor detail that was adjusting the rail tracks in anticipation
of their arrival.

> In the air . . . you could already sense the danger. . . . Everyday, we
> went to the railroad, to reroute the tracks from Auschwitz straight
> to Birkenau. When the job was done, the Jews of Hungary were sent
> straight from the train to the crematoriums. Every day we heard that
> they were coming. In April 1944 a transport of Greek Jews from
> the vicinity of Athens arrived. This transport came in just before the
> transports of the Hungarian Jews. I remember that most of the
> Greeks who arrived at that time were sent to work in the Sonderkom-
> mando and the crematoriums. But just then, the massive transports
> of the Hungarian Jews arrived . . . [and] the transport of the Jews of
> Athens was swallowed within its mass.[75]

The last transport from Greece—from Corfu—arrived in Auschwitz in
June 1944. Almost all on board were gassed, with no selection.[76] In mid-
August, a final transport of Sephardim, seventeen hundred Jews from
Rhodes, arrived from Athens, along with about ninety Athenians who
had been rounded up in the previous months.[77] The last gassings at the
camp took place two months later, on August 18, killing two thousand
Jewish inmates brought from Terezin.[78] The first, which killed fifteen hundred Polish Jews from Sosnowicz, had taken place more than two years
before, on May 12, 1942.

The Thursday, February 1, 1945, lead article in *Kathimerini* has as its
headline "*Tría Hrónia*," three years. It is a sad, angry, and moving inven-
tory of all that the Greeks had suffered during the occupation. "LET US
REBUILD OUR HOUSE, GREECE. Let us remake our state. Dictatorship, kingdom, democracy, laocracy . . . it doesn't matter. Just a state.
With . . . justice . . . and [built on] earth that will be watered with holy
water."[79] The article recalls the hunger, death squads, and countless
Greeks whose friends, family, and neighbors had been brutally killed. It
bitterly recounts the opportunistic meddlings of the great powers, and
Greece's utter isolation in her time of need. It says nothing, however, of
her Jews, gone as if sunk beneath the waves.

At the beginning of World War II, there were over seventy thousand
Greek Jews. At the end, there were about ten thousand. In terms of percentages, this reflects one of the highest destruction rates in all of Europe.
As was the case with virtually all other aspects of Greek Jewish history,
its course was determined in large part by local circumstances. For the
first time, though, Greek Jews—in a perverse irony—became within the
framework of the Final Solution one unified national entity. No longer
were there Apulians or Sephardim, Ashkenazim or Romaniotes; no longer

were there Jews of Old or New Greece. There were simply Jews, targeted for death. For years, the outside gaze of Christian Greece had reinforced their status as Jews. But in the camps, the outside gaze of others—Nazi guards and Jews, but mostly Jews—reinforced their status as Greeks. In Auschwitz, which destroyed so many Greek Jews, Jews from Greece felt their Greekness more acutely and poignantly than ever before. There "they felt like Greeks, and they died as Greeks."[80]

Everyone in the camp was fascinated by the Greeks—German officers and scientists, and Ashkenazic Jewish inmates alike. While the Germans measured their heads, searching for signs of their genetic superiority, and recruited them for the most gruesome work details, their fellow inmates marveled at their mysterious silence, handsome physiques, and overwhelming sense of solidarity even as they looked down at them as inferior Sephardim. The argot of the camps was heavily inflected by Greeks, and the sounds of the camp—the singing, speech patterns, and cheers of the crowds who gathered to watch the Greek boxers—were strongly shaped by their presence. Under the gaze of these outsiders, Jews from Jannina, Salonika, Thrace, Athens, and Corfu—from all the far-flung corners of the country—became one national conglomerate. They became Greeks. The process would be furthered in the decades after the war, in a quite different site of heterogeneous mass Jewish cohabitation: Palestine/Israel.

Chapter 9

TRYING TO FIND HOME:

JEWS IN POSTWAR GREECE

> An almost hysterical urgency for help . . . prevailed
> [in Greece] at the end of the war when Jews began
> returning from their hiding places.
> —American Joint Distribution Committee
> meeting minutes, May 13, 1947

DURING THE 1920s and 1930s, a fledging distinct Greek Jewish identity had emerged. The youngest generation particularly of Salonikan Jews, those who came of age under Venizélos and Metaxas, were culturally and ideologically more Hellenized than any generation before them. Greek Jews increasingly occupied diverse points on the political spectrum, rather than voting as a bloc, as had traditionally been the case. Almost all spoke Greek as well as Ladino. With increased secularization, and especially the loosening of dietary law and Sabbath observance, the social worlds of Jews and Christians were becoming intertwined. Counterhistorical speculation is tempting: Might Salonika have become the premier modern Jewish world city? Would Jews have become a Greek political and cultural elite? Would Greek nationalism have developed more secular or multicultural strands, or would Greek Jews only become fully Greek through assimilation, by essentially losing their Jewishness? All possibilities were stopped in their tracks by the Nazis.

If in Greece the term Greek Jew had meant little until the decades before the war, after the war it scarcely had a referent. Paradoxically, although not coincidentally, the Greek state itself used the category for the first time only in the wake of the near decimation of Greece's Jewish population. It is only in the postwar period that Greece for the first time created a national Greek Jewish representative body. Under other circumstances, this would have been a vitally important development in Greek and Greek Jewish history. But this body represented only the tiny *she'erit* [remnant] of Greek Jewry that remained: about eight thousand people, about 10 percent of the prewar Jewish population, and about one-tenth of a percent of Greece's total population.

Not since the close of the Greek War of Independence had Jews made up such a tiny fraction of the Greek populace. In the wake of independence the Greek state had felt the triple pressures of international intervention, the consolidation of a distinctly Christian form of national identity, and a society divided between those who had fought for independence (largely autochthons) and a new political elite. Jews, miniscule in number, were regarded as traitors for their collusion with the Ottomans. In the years after the Second World War, similar factors were again at play. Greece in the 1940s was characterized by the increasing involvement in its domestic politics of Great Britain and the United States, which saw it as a critical player in the emergent cold war. A reassertion of Orthodox Christianity as the bedrock of the nation worked hand in hand with anti-Communist sentiment. And one whole flank of the Greek resistance, Greek Communists, were exiled from Greek political life—executed, imprisoned, or (and this was particularly the case with Greek Jewish Communists) driven out of the country altogether as anti-Greek traitors.

While on an institutional level Greek Jews enjoyed more official state recognition than ever before, in terms of Greek Jewish history the reality of the postwar era was that the category Greek Jew was a phantom one, assigned to a people all but extinguished within the borders of Greece itself. Among the few who had survived, there was little sense of collectivity, and the fragmentary community that remained was divided along multiple lines. Immediately after the liberation, the Jewish Communal Councils of Athens and Salonika reaffirmed the Greekness of Greece's Jews, presenting a Greek flag to the Athens government and thanking the Christian population for its assistance during the occupation.[1] But the collectivity represented by such gestures was an elusive one. After the war the frantic need to reconstruct individual lives, crushing poverty, and the lure of emigration dominated the lives of survivors. Half of all camp survivors who returned to Greece stayed only briefly, moving on quickly to Palestine/Israel, Africa, or when possible, the United States. Hundreds went to Palestine directly from Europe's displaced persons camps, after brief stopovers in Spain and Morocco.[2] The consolidated national identity they had acquired in the camps largely went with them, and was reinforced, if recast, in new diasporic contexts.

SURVIVORS

Just after the liberation there were a few thousand Jews in Athens, who had hidden with Christian families during the occupation. Another three or four thousand were in the north, with the partisans. In Salonika, few Jews had survived in hiding.[3] Aside from members of the resistance, the

Fig. 9.1. Jews hidden in Christian homes, collected after the war by the welfare squad, 1945–46, Greece. Permission Film and Photo Archive, Yad Vashem, Jerusalem.

Jews who returned to their hometowns after the war were either survivors of the camps or individuals who had gone into hiding with Christian families during the course of the war. On arrival, they were confronted with a host of problems. Jews from the camps were poorly treated by Jews who had survived the war in hiding, who could neither believe nor understand what had happened in the concentration camps. Members of the resistance—or anyone with ties of any sort to Russia—were tarred as Communists. Many confronted the resentment and impassivity of Greek Christians who thought they were gone forever and had laid claim to their property. When asked to return the property, some Christians went so far as to express regret that not all the Jews had been killed. All of the survivors were depressed and stunned by the magnitude of the destruction.

The physical as well as human landscape was unrecognizable, bereft of familiar markers and inhabited by strangers. Entire neighborhoods had disappeared, and virtually no known faces were to be seen. The Salonika harbor was littered with the hulls of sunken ships; whole towns had been gutted by the occupying forces. Towns once full of life had been transformed. In late May 1945, the U.S. Navy ship *Park Holland* docked in Patras. Among the armed guard unit on board was Louis Levy, signalman third class. Levy's parents were Ioanniote Jews who had emigrated to New York in the 1920s; Levy had never been to Greece before. Unable to resist the pull of his ancestral hometown, he went AWOL and risked court-martial to travel secretly to Jannina. Levy's account of his trip describes devastated villages and virtually impassable roads.[4]

For survivors of the camps, a further cause for alienation was the horror with which those who'd spent the war in hiding regarded them. In March 1945, Leon Batis, an Athenian, returned to Greece. The first Greek survivor of the camps to make his way back to the country, he left Auschwitz in December 1944, heading south on foot and by rail. When he arrived in Salonika on March 15, word quickly spread among the few hundred Jews who had survived in the city that a fellow Greek Jew had returned. At the time, even at the state level there was little clarity as to what had become of Greece's Jews; in the opening months of 1945, the British, Greek, and Spanish governments wrote feverish memos in the effort to learn.[5] Had they been sent from the camps to Africa? Were they still in northern Europe? Had they died of thirst and hypothermia en route to Poland?[6] In January 1945, one of the few Salonikan survivors had written desperately to the British Embassy, "If you feel that you can do anything towards locating the present whereabouts of the Jewish population deported from Salonika, you would be rendering us a sacred favour."[7] Batis, as the first eyewitness to return to the city, was besieged with confused questions from people who hoped to hear news of their long-departed friends and relatives. All wanted to know: Where had everyone been taken? What had happened to them? He sat for hours drinking ouzo and recounting his story. "We sat silently around Batis, eight Salonikan Jews, and we listened avidly to the account of the first, who had gone away and come back to Greece." He told them of the transports, the camps, the crematoriums and Sonderkommando, and those who had to pull out the gold teeth of the dead. "It was unbelievable, and Batis was right to think that we'd take him for a macabre hallucinator." All the same, his interlocutors had the sick feeling that perhaps he was telling the truth. Less than three weeks later, six more survivors arrived.[8] Gradually, those who had eluded deportation began to realize the magnitude of what they had escaped.[9]

Ovadia Baruch was among the first Salonikan survivors to return home to their city, traveling on foot through Yugoslavia with a small group of fellow Greeks. The group arrived late at night and were sent to the local *han*.

Only in the morning did we make our way to the Jewish community [those that had come out of hiding]. And there, a whole new story began. No one even wanted to look at us. For some reason, they thought that we had killed all the Jews! They sent us to sleep in the synagogue. . . . For three weeks, I slept there on a bench with no blankets. This is the treatment I got in my own homeland from our Jewish brethren. Slowly, slowly, more survivors arrived, and the truth about Auschwitz was revealed.[10]

Many returnees were interrogated in the Pavlos Melas prison in Salonika.[11] They felt that they were being treated as potential criminals when they should have been welcomed as survivors of German brutality.

For those who had spent the war in hiding, some in the very towns in which they had grown up, the experience of seeing the first few survivors trickle back in 1945 was an uncanny one. While they had heard rumors, for the most part they had no idea what had become of their departed neighbors. They were suspicious, surprised, and confused when some started to return. Representatives of the American Joint Distribution Committee (AJDC) stationed in Salonika commented on the psychologically fraught interactions between camp survivors and those who had gone into hiding, and the tensions produced by the gradations of suffering that different categories of survivor had endured. In postwar Greece, the tenuous national solidarity that had characterized Greek Jews before the war and the tremendous solidarity experienced by Greek Jews in Auschwitz were both fractured by the divisions created by the very different experiences individual Jews had had during the war.

We had a difficult time getting Greek Jews who remained in hiding to take an interest in their brothers who came back from concentration camps. They have all suffered, but you cannot compare a person's lot who survived in the mountains with somebody who came back from the crematory.

And it appeared cold and heartless that they would not take an interest in those who came back. But we began to understand why. You see, 63,000 had been deported; two thousand came back . . . and in the main they were not the relatives of those who survived in hiding. In the main, the relatives of those who survived in hiding were destroyed. And then a Greek Jew who survived in hiding met a Jew back from the crematory or concentration camp or both. And he said, "Do you know my mother, my sister, my child?" And he was informed, yes, such and such a person had died in the crematorium, in which possibly the person who returned had worked.

There was the feeling visualized, that the people who came back were not as good as those who were destroyed. Some people went so far as to feel that they [the returnees] had destroyed their loved ones.[12]

Returning Jews around the country had similar experiences. Freida Kobo, a Salonikan Auschwitz survivor, was flown to Athens. "No one took care of us. I remember terrible incidents—we weren't even allowed in the synagogue, and it was the High Holy Days." The Athenian Jews who had not experienced life in the camps "treated us as second-class human beings, because we didn't reserve seats . . . this is what is hard for us to understand. Even in Greece, they did not welcome us."[13] In Jannina, the earliest survivors to return from the camps arrived in town in early 1945, but wouldn't say much about where they had been or what had happened. "One day the first survivor came. Traumatized, skinny, with a number on his arm, written indelibly, 77.825. 'Don't worry,' he told us, 'everything is fine and they'll return.'" As others gradually arrived, those who had escaped deportation heard about the horrors of the camps, the selections, and the murder of more than 88 percent of the members of their community. "We understood that there wasn't anyone else [alive] to return."[14]

In some cases, lurking behind the hostility that different categories of survivors had for one another was the specter of collaboration. Some survivors were regarded as collaborators, either because they were related to people who had collaborated or because they had behaved in ways viewed as contrary to community solidarity. Sometimes, the mere fact of survival was taken as evidence of collaboration. Suspected individuals were shunned by other survivors, and many emigrated in the effort to escape social ostracism.[15] In contrast to the Greek government, the Jewish community was aggressive in the pursuit of collaborators.[16] Prominent trials in 1946 and 1947 riveted the community, but also heightened the paranoia that prevailed between different groups of survivors.[17]

For their part, the Jews who had survived the war in hiding did not know what to do in the face of the news of what had happened in the camps. As the few returning deportees recounted their stories, disbelief turned to horror and then guilt. Roza Asser-Pardo, who as a young girl survived deportation by hiding with Christians in Salonika, recounts that "[for] many years we were ashamed to meet [each other] and talk about how we, too, we who had been in hiding, had suffered, even if we didn't reach the point of death."[18]

While Jewish agencies and the Greek government itself developed institutions that treated Greek Jews as a national collective, among Greek Jews themselves there were tremendous divisions. Again, the rift between those who had survived in the camps and those who had survived in hiding was perhaps the most striking. (Between a third and a quarter of all survivors had survived the camps; the remainder had survived in hiding.) Converts to Christianity made up a third group. During the occupation, large numbers of Jews, particularly in Athens, had been given baptismal certificates. Others were hidden with Greek Christian families; some underwent baptism and conversion. Children who came of age during the war had little recollection of their former, Jewish lives; many adult converts felt that there was no hope in a return to Judaism. The Tel Aviv paper *haMashkif* (Observer) in summer 1945 commented on the "extensive conversion to Christianity of Greek Jews, who proclaim their desire not only to stay in Greece but also not to return to the Jewish religion. . . . [They] state quite clearly that they will never return to Judaism . . . convinced that the tragic fate of the Jews will never change." While huge efforts were under way to bring as many Greek Jews as possible to Palestine, the article concluded that in the case of such resolute converts, no pressure should be brought to bear. As the Greek consul summarized, "The newspaper [article points out that] . . . it would be immoral to spit now in the wells whose water had saved them from death at the most tragic moment in their lives."[19] Converts tended to have little connection with the Greek Jewish survivor community; indeed, most did not consider themselves part of the Greek Jewish community at all.[20]

FAMILY FRAGMENTATION AND FEMALE SURVIVORS

Greek Jewish cohesion was also shattered by the fact that virtually no family had remained intact. Single Jewish women with neither birth family nor spouse had been unheard of before the war, and female survivors were particularly ill equipped to negotiate the postwar environment. The young single Jewish woman was a new phenomenon, perceived by Jewish leaders as an especially potent threat to the cohesion of the community. In late 1945, *Israilitikón Víma*, a newly constituted Greek Jewish newspaper in Salonika, published an article focusing on "the Jewish woman, for whom the post-war situation has become so tragic because of the economic and ideological crisis that now dominates our country." The article expressed the deep concerns of the community's leaders about the future of Greek Jewish women, given their economically and morally degraded state. The latter was of particular concern. "If things continue in the direction they've taken, despite all our efforts many of our young girls will take a bad road." While men had lost their economic standing, families, and homes, young women had lost "the most important [thing] of all":

their innocence and ethics. Many had been raped, widowed at a young age, and torn from their families. Almost none had male oversight, and few had prospects of marriage.[21] Some had worked in what in Auschwitz parlance had been called the Pouf Kommando, which performed sexual favors for the German guards.[22] In Salonika after the war, some pregnant women were suspected of carrying half-German children.[23]

While religious life had revolved around men, its ethical core had been the community's women. The reconstruction of the Jewish community after the war was undermined by the dramatic transformation that the experiences of the war had brought to the lives of Greek Jewish women. The lived markers of Jewish life—kashrut, Sabbath observance, and tradition (*minhag*) had been sustained largely within the family. Within two years of the war's end, almost no Greek Jews kept kosher any longer, Sabbath observance was nonexistent, few communities had a rabbi, and almost none had a mohel (ritual circumciser) or *shohet* [kosher butcher]. Religious education was unavailable.[24] Synagogues were unusable.[25]

Organizations attempted to target these problems by focusing on the plight of women, "the most precious heritage remaining to us."[26] But rehabilitating women was not the same as rehabilitating the community. Indeed, rehabilitation actually meant providing women with the necessary skills to lead an independent life and the wherewithal to leave the country. The Athens Home for Jewish Girls, established in July 1946, was sponsored by the National Council of Jewish Women. The home cared for about sixty-five young women before ceasing operations in December 1949. Almost half of them ultimately immigrated to the United States; several married U.S. GIs.[27]

Much as many women had rushed to marry immediately before deportation in the hopes that it would bring some security, those who were able now hastened to marry in the wake of liberation.[28] Marriage was security, a return to normalcy, and a way to establish adulthood. As one woman who spent the occupation in hiding commented in 2001, "I didn't have a childhood. I grew up too fast. We had to grow up. I married too young."[29] For many women, the difficulties of being alone while trying to make property claims, reestablish old lives, and reclaim normalcy were too great to be undertaken alone, particularly for those who had no prior experience in dealing with financial and bureaucratic matters. Marriage was a bulwark. But while before the war it had been an institution that drew the Greek Jewish community together, in the postwar period it was another factor in fragmentation, as women married men from distant towns or outside Judaism altogether in response to demographic and economic pressures. Many married in the Zionist *Hakhshara* (training camps for emigrants to Palestine, a number of which were in operation in

Fig. 9.2. Ten couples getting married in the Zionist Hakhshara, 1946, Athens, Greece. Permission Film and Photo Archive, Yad Vashem, Jerusalem.

Greece) in anticipation of leaving the country altogether. In Jannina, the overwhelming majority of survivors were women. In Kavála, there were almost no Jewish women at all.[30] And even if women could find a groom, they usually couldn't find enough money for a dowry—"and you don't just get married in Greece unless you have a dowry."[31]

PROPERTY RESTITUTION

A major obstacle to community reconstruction was the utter devastation of Greece as a whole. As Roosevelt declared, "No country except Poland has been reduced to such bitter destitution."[32] As early as 1942, hundreds of villages had been abandoned or destroyed, almost all the sheep and goats had been killed off for food, and agricultural production had been halted.[33] By the war's end, inflation had reached heights unprecedented in the history of modern currency. Of Greece's eight railway lines, six were rendered completely useless, the two principle ports (Piraeus and Salonika) both needed years of repairs, the Corinth canal was "completely out of commission," and all of Greece's sixteen thousand kilometers of paved road needed repair. Seventy percent of the phone lines were destroyed. Out of a total population of 7.2 million, 1.2 million were homeless.[34] Greek Jews were in competition with millions of needy Greeks; since organizations such as the United Nations Relief and Rehabilitation

Administration (UNRRA) had resolved to tend to the needy "without discrimination," Greek Jews were to receive no special treatment as the difficult process of reconstruction began.[35] As the AJDC acknowledged, this was not a reflection of anti-Semitism but rather only of the fact that the country as a whole was utterly devastated. At the same time, international refugees—Dutch and Belgian, particularly—were coming into the country across the northern frontier at the rate of about one hundred per week.[36] Greece was the major land route for illegal immigration to Palestine and was under pressure from the British to do all it could to stop it. The Greek government was also anxious that it not be burdened with the economic cost of taking care of refugees and warned of "the danger of their being registered as members of [Greece's] . . . Jewish Community"— and as such, registered as Greek citizens.[37] Greece formally refused all "Jews of non-Greek origin . . . entry into Greece," although exceptions were made.[38]

But in one realm Jews were at a huge disadvantage vis-à-vis other Greeks. Most had been propertied before the war; after it, none was the legal owner of anything. Survivors returned from the camps or emerged from hiding to find their homes occupied by Greek Christians, and their furniture and possessions sold or stolen. Property restitution claims emerged as a central component of the complicated relationship between Greek Jews and Greek Christians. The AJDC reported that anti-Semitic incidents almost invariably had "economic motives" behind them, most related to contested property.[39] Anti-Semitic incidents were reported with far greater frequency in places where large amounts of Jewish property were contested, such as Corfu and Salonika, than in those where Jewish communities had been poorer, like Kastoria and Florina.[40]

Property claims were so central to postwar Greek Jewish life that they provided the initial impetus for the creation of Greece's first legal corporate Jewish entity. In 1945, the Central Board of Jewish Communities (Kentrikón Israilitikón Symvoúlion, or KIS) was established as part of the state apparatus of the kingdom of Greece. Its initial purpose was to assist the reintegration of survivors into Greek society and help with property reclamation; today it operates under the Ministry of Education and Religion, and is the corporate body via which Greece's various Jewish communities interact with the Greek government and foreign organizations.

Greece was the first country in Europe to promulgate laws to assist Jews trying reclaim property.[41] The laws' enforceability, however, was limited. Just before the liberation of Greece, M. Alexandros Svolos, the president of the Committee of National Liberation and the newly appointed minister of finance in the Hellenic government, had gone to Cairo, where he reported to the exiled government on conditions in Greece. Emphasizing the particularly dire straits of the country's Jews, he declared that one of his first acts in office would be to facilitate the restoration of all Greek homes and property.[42] As soon as the occupation ended, the government established the Special Department for Jewish Affairs, within the Ministry of the Interior, which was to determine the number of Greek Jews who had been deported, make arrangements for orphans and widows, find and prosecute collaborators, and help with property restitution. The move was met with enthusiasm by international Jewish organizations.[43] But the work of property restitution would prove far easier said than done.

In the wake of the deportations, almost all Jewish property had been sold in sham transactions, stolen outright, or taken possession of by the Greek government under laws 1977/44 and 1180/44. By law, the Department for the Management of Jewish Property was to warehouse the possessions of deportees; taxes levied on the impounded property were to cover the department's expenses. On paper, the system that had been put in place looked efficient.[44] In practice, though, it was in most instances meaningless. In Salonika, the Greek committee established as a clearinghouse for Jewish property during the deportations had been completely overrun by the German authorities, who gave collaborators first dibs on the most choice properties and possessions. In Jannina most notably, but in almost all other towns as well, many members of the Christian population went into Jewish districts immediately after the Jews' departure and helped themselves to whatever they wanted. On Corfu, the Germans handed all property over to the Greek government almost immediately, which in turn sold it to Christians.

Soon after liberation, in November 1944, King George II rescinded laws 1977/44 and 1180/44, and put in place various measures for the restitution of property.[45] By the following summer, the World Zionist Conference was informed that the Greek government had withdrawn all claims to Jewish property, and that the property was to revert immediately to the former owners.[46] All property that had no claimants—a frequent circumstance, as in many cases entire families had been killed—would not become the property of the state (as was normally the situation with cases of individuals' dying intestate) but would be turned over to the Jewish community for charitable purposes under law 846/1946.[47] The Agency for the Care and Rehabilitation of Greek Jews was established to administer the law and coordinate Jewish aid within the country.[48]

The state's position on Jewish property was forward thinking, and the state was involved at several levels in trying to assist Jewish survivors with their claims. The reality on the ground, however, was that most Jews were unable to overcome the obstacles posed by the attempt to reclaim property. In some cases, Jewish claimants were met with an inhuman hostility. The experience of one Salonikan Jew who returned to his home after the liberation of Auschwitz was extreme but not unique. He told the Christian family that had occupied his home, "'I am back. My wife and children were cremated. I want one room in my house.' He was rebuffed with, 'Get out of here. I am sorry that you are not cremated also.'"[49] Such cruelty was not characteristic of all cases, but endless red tape and wranglings were. The experience of David Mordoch, an Auschwitz survivor from Salonika, was typical:

> In 1946 we returned to Greece. . . . I didn't find anybody. I went to live in the house that had belonged to my father, but I was thrown out. . . . Until 1949 I lived in my parents' house, after a struggle I had with a Greek who lived in the house and didn't want to give it to me because, he said, it belonged to my parents, [not me]. And since they did not return from the camps, I didn't have the right to live there. I remember the terrible pain [of this]. This was the house I grew up in, from this house I was sent to the camps. And here someone was preventing me from returning to it. Finally I managed to invade it, and finally to drive that Greek away.[50]

But when Mordoch was drafted into the Greek army in 1949, to fight in the Greek Civil War (1946–49), he gave up citizenship and left for good.

In Salonika a vast quantity of valuable property had been left behind by the deportees, and the Salonikan community was consumed with the restitution process. Over two thousand Jewish businesses had been left vacant in 1943; by late 1945, returnees had regained only 8 to 9 percent. Over ten thousand homes had been vacated, yet returning Jews found that once they got to Salonika "after surviving Hitler's crematoria," they still "continued to live under the same sorts of circumstances as in the camps," sleeping on cement floors, without blankets, in ad hoc shelters.[51] They felt like "refugees in their own city."[52] In Jannina, 170 survivors returned in 1945–110 from the camps, and the rest from hiding in Athens or the mountains. Six hundred Jewish homes in Jannina had been vacated, but survivors were crammed into ten houses, where they lived under intolerable conditions. "Owners who returned naked, exhausted, and desolate . . . managed with difficulty to reclaim their family homes [and] were considered lucky if they were able to get just a little corner where they could rest their exhausted bodies." The local authorities remained largely impassive in the face of the suffering of the returnees.[53] By early 1946, only ten of Jannina's shops were under Jewish ownership; all had been emptied of their wares just after the deportations. Two hundred and twenty Jewish-owned sewing machines had been turned over to Christians after the deportations. Jannina's Jewish community wanted to get them back so that Jewish women would have some means of making a living. Litigation over the sewing machines alone took months and was concluded unsuccessfully.[54]

While the law seemed to provide for rapid restitution, there was a "gaping chasm" between it and reality.[55] Children whose parents had died in the camps had to spend months proving their claims to family property, while Christians who had benefited from the Jews' forced departure made efforts to slow down the process and keep property for themselves. A few years after the war some had managed to reclaim property, but slowly and bit by bit. "They usually get back first legal title, then they get back one room, a bare room."[56] The AJDC complained that "the usual andante tempo of Greek Governmental affairs is naturally retarded to adagio when money is involved."[57] In court, returnees more often than not lost their cases. When they won, judges acted as if they had treated the plaintiffs with generosity and charity, rather than with justice.[58]

The bitter process of property reclamation made many Jews feel as if they were unwelcome strangers in their homeland. Veterans of the Albanian front complained bitterly that after making so many sacrifices for Greece, they had returned home only to find that the suffering inflicted by Nazism was being continued in their own land. At the same time, it had a divisive effect on the Greek Jewish community as well, underscoring the difference between Salonikans, who had tended to be relatively wealthy before the war, and Jews from other parts of the country. Outside observers commented on the superior attitude that the once-wealthy Salonikan Jews had toward their poorer coreligionists, who tended to be concentrated in Athens.[59] Finally, the looming matter of disputed property emerged as a factor in emigration: Jews who had owned property in Greece tended to maintain ties to the country for longer than those who had not. The loss of property was in many instances a decisive factor in the decision to leave Greece.

INTERNATIONAL AND DOMESTIC AID

As Violetta Hionidou has recently documented, a propaganda and politics of relief—in which the Greek and British governments, international agencies, and the occupying forces participated—had emerged during the occupation as a striking feature of the domestic landscape. Among its features were a dramatic expansion of preexisting Greek welfare structures, and a jockeying for position between international and domestic agencies.[60] On liberation, the politics of relief and restitution had particular implications for Greek Jews. Within Greece, KIS became the permanent institution via which Greek Jewish interests were represented. During the 1940s, it devoted its time to property restitution and the rebuilding of Jewish life, sending young Greek Jewish men abroad for rabbinic training, sending religious books to the Jewish communities around the country, and establishing some Jewish education in Greece. Further, it monitored the whereabouts of known war criminals and pushed for their prosecution.[61] (By the 1950s, it had also emerged as a sort of transnational entity, serving as a liaison between Greek Jews in Greece and those who had emigrated to Israel.)

Various international Jewish organizations were active in postwar reconstruction efforts. There was conflict between different organizations, and confusion as to their legitimacy and efficacy due to the chaotic political situation that prevailed in Greece, with dizzyingly swift regime change and no clear sense of who was in charge. "A commitment may be made by one minister who does not remain long enough to fulfill it, and the matter has to be started all over again with his successor who may also be out of office before anything is done."[62] Greek Jews themselves weren't sure of the chain of command, and in the effort to circumvent the confusion made direct appeals for help to the Supreme Headquarters of the Allied Expeditionary Forces and other outside entities that had no jurisdiction whatsoever over Greek Jewish matters.[63] International representatives didn't know with whom they were meant to negotiate. As one lamented, "There are many Greek governments. There was a short period when there was no government."[64]

Internally, the mandate of many aid organizations was unclear, with activities divided between the contrary goals of helping reestablish Greek Jewish community and fostering emigration to Palestine, and they drew Greek Jews into an international Jewish orbit as much as carved out a distinctly Greek one. Dominant in this regard was the AJDC, which was active in Greece from immediately after the war (although its legal status, like that of other Jewish organizations, was initially in doubt, since aid was supposed to be rendered "without prejudice," meaning that the needs of no one ethnic or religious group were to be put above those of another). The AJDC remained operational there until 1950, advocating for the well-being of Greek Jews on a number of levels, coordinating health care for tuberculosis sufferers, providing funds for the support of unemployed Jews, and helping the various communities communicate with the Athens government.[65] It also organized a cooperative loan bank to assist Jewish survivors who wanted to establish independent businesses.[66]

Almost from the start, there was friction and even outright hostility between the AJDC, KIS, the Jewish community of Salonika, and other international organizations, notably UNRRA. At issue was the AJDC's treatment of Greek Jews compared to others. As Greece was a favored route for access to Palestine, hundreds of international Jews arrived in the country monthly. There they turned to the AJDC for help. It "looked after Jews in Greece irrespective of whether these were residents or refugees arriving across the northern frontier."[67] Thus, its efforts were divided between helping international Jewish refugees get to Palestine and helping Greek Jews reestablish community in Greece. Over time, the two mandates would become confused, and increasingly the AJDC would be seen as a clearinghouse for Greek Jewish emigration to Palestine, so much so that in November 1945, KIS sent a telegram to the Foreign Office in London to say that it "recognize[d the] Jewish Agency [for Palestine] as sole representative body entitled to speak on behalf of the Jewish people as a whole in all matters affecting Palestine."[68] The Greek government, too, was hostile toward the AJDC, as it was toward many international organizations, particularly after the start of the Greek Civil War, when outside assistance was feared as a front for leftist intervention.[69]

When it came to choosing between international refugees and Greek Jewry, Greek Jews felt that the AJDC regarded their needs as the lower priority. When it pulled its representative out of Salonika in 1947, the Jewish community there argued that the AJDC did not understand its needs and was stingy with assistance.[70] Its representatives had a "false sense of pride and haughtiness," and were "completely indifferent to the needs of the people who come to them for assistance."[71] When the AJDC countered that changes in Greek law would soon bring large rental revenues to the community, so outside assistance would no longer be necessary, the Jewish community of Salonika wrote back angrily, "As for . . . so-called 'Jewish wealth,' the Government will never return any Jewish properties to the community."[72] Survivors were also furious at what they perceived to be the AJDC's complicity in handing over Jewish property to the Greek government. As an increasing number of Jews left the country altogether, the question of what should become of community-owned property became urgent. Before its departure, the AJDC worked to negotiate the sale of Salonika's Hirsch Hospital to the government of Greece, which had requisitioned it since demographics "no longer justify use of the hospital by the Jewish community alone."[73]

For its part, the AJDC complained that Greek survivors didn't grasp that the end goal of charitable aid was self-sufficiency, and levied the accusation that international aid had rendered Greek Jews passive and overly reliant on handouts. "For many people the emergency period is over, but they see no objections to the Joint's continuing to pour in its goods. Bitter antagonisms are aroused when they are informed . . . that they are no longer entitled to relief."[74] The survivors were too needy, demanding, and unwilling to take control of their own destinies. "Persecution does not make better people, unfortunately," commented one AJDC official after spending fourteen months in Greece.[75] Its officials also complained of the "general suspicion evinced by the [Greek] government towards all foreign agencies which did not come directly under its control and supervisions."[76]

Other international organizations, notably the Sephardic Brotherhood of America, also sent representatives to Greece and advocated for survivors.[77] The Argentinian Committee on Behalf of Jewish Victims sent financial assistance, as did the Jewish communities of Egypt and South Africa.[78] The Jannina Relief Fund raised $50,000 for the Jews of Epirus, paid the dowries of women who wished to marry, and organized vocational training for single Jewish women in the region.[79] To some extent UNRRA, too, worked on behalf of Jews, particularly through its Palestinian Jewish Relief Unit. Most UNRRA resources were committed to general reconstruction projects rather than specifically earmarked for Holocaust survivors. Like KIS and the Jewish community of Salonika, UNRRA became embroiled in conflicts with the AJDC, mainly over the AJDC's encouragement of illegal emigration of Jews from Greece to Palestine.[80] At the same time, UNRRA itself was accused by the British and Greek governments of assisting the Jewish Council for Palestine's efforts to encourage emigration to Palestine—charges that UNRRA vociferously denied.[81]

The Jewish National Council for Palestine established offices in Athens in 1945, and had branches around the country. The Jewish National Council's primary mandate, to help Jewish refugees immigrate to Palestine, was doubly problematic, as it brought unwanted Jewish refugees into Greece and fostered illegal emigration out of it. It came into frequent conflict with KIS, the Jewish community of Salonika, and the Greek government, although KIS favored it over the AJDC. The council repeatedly petitioned the Greek Ministry of Public Security, asking permission to bring international Jewish refugees into Greece.[82] After allowing one group of refugees into the country, the Greek government, largely at the urging of the British, subsequently denied such requests.[83] There was no money to spare to feed and care for refugees. More pointedly, the Greek government expressed concern that figures for Greece's Jewish population not get overinflated. Jewish refugees might blend in too easily in Salonika and settle in Greece.[84] At the same time, though, the government objected to the Jewish National Council for Palestine's meddling in the affairs of Greek Jewish citizens and condemned the efforts it was making to get Greek Jews to move to Palestine.[85] At the level of the local authorities, however, there was little compliance with the government's official stance, and in a number of regions, Greek authorities turned a blind eye to the council's activities. From Corfu, for example, the British consul wrote, "My impression is that the Greek Authorities here will not be too energetic in restraining the Jews from leaving . . . as apart from the desire to be rid of them there is a certain amount of rivalry over the division of property belonging to non-returned Jews."[86] The British and Greek governments tried to monitor the activities of the council, setting up surveillance of its agents in Patras and other towns on the coast.[87] By summer 1946, up to half of all Jewish relief teams working in Greece were, as the British consul general stationed in Salonika put it, "act[ing] as cover for Zionist schemes."[88]

In the midst of all this confusion, where were the Greek Jews themselves? The conflicts between the AJDC, the Greek government, and KIS reflect, to a large extent, the slippery meaning of Greek Jews. Were Greek Jewish properties Greek or were they Jewish? Were Greek Jews meant to be reestablished as a community within Greece or should assistance instead take the form of relocation to Palestine? Were Jews within Greece deserving of more help than any other Greek? Were Greek Jews a subset of world Jewry or Greek citizenry? These questions were as much on the minds of Greek Jews themselves as anyone else. And within the community, too, there was debate as to their answers. The Zionist Federation of Greece, for example, not surprisingly was fiercely opposed to British efforts to stop emigration from Greece to Palestine. The Jewish community of Athens adopted a resolution demanding "that the doors of Palestine ... be opened wide to Jewish immigration without restriction," and proclaiming its "complete solidarity with the Jewish Agency for Palestine and the Palestinian Yishouv."[89] Athens, where up to half of all survivors had settled, was a center for Zionism and Jewish emigration. KIS's efforts, which centered on Salonika but extended throughout the country, were focused on stocktaking: gathering statistical data on the number of survivors from each Greek Jewish community and mapping out a plan for the reconstruction of Jewish life.[90] As a recognized entity of the Greek state, KIS was able to gain corporate benefits for the Jewish community that in the prewar period were negotiable only on local or ad hoc bases; through KIS, Greek Jews were brought closer than ever into the administrative mainstream of Greek society. An appeal to the Department of Education, for instance, resulted in the incorporation of Hebrew language and Jewish history into the curricula of schools where Jewish students were enrolled.[91] Its leadership, notably Asher Moissis (KIS president from 1944 to 1949), also advocated for the translation of Hebrew texts into Greek with the goal of keeping increasingly Hellenized Greek Jews in touch with Judaism. Religious Jews (of whom there were few at war's end) found it virtually impossible to live an observant life in Greece. Zionist Socialists saw the war as confirmation of the absolute necessity of Palestine. Secularized, Hellenized Jews were more prone to remain in Greece. The fact of being Greek and Jewish was not, on its own, a unifying factor powerful enough to hold the tiny surviving community together.

THE CIVIL WAR AND EMIGRATION TO PALESTINE

Up to half of all Greek Jews who survived World War II did so with the assistance of the leftist resistance. Many were themselves members of the resistance; others had spent the German occupation hiding in the mountainous north of the country with the help of partisans. The Communist EAM had been the only Greek political organization that systematically arranged for the protection of Jews during the occupation, and virtually every Jew who escaped deportation had some contact with it. Already well before the war many Jews had joined the Greek Left; the leftist parties that had held increased appeal for Greek minorities starting at the turn of the century were the mainstays of the EAM.

The huge importance of the Greek Left to the survival of Greek Jewry was to become yet another complicated irony in their torturous history. In the postwar period, Greek political life was dominated by the Civil War (1946–49) and the purge of the Left. Jews who had found in the resistance a potent vehicle for national sentiment and loyalty to Greece discovered that their politics now made them suspect as traitors instead. All survivors, many only recently freed from the death camps, were faced with Europe's bloodiest civil conflict, into which all male citizens were drafted in 1946. The tremendous pressure brought to bear on Greek Jews by the postwar political environment proved arguably to be the single greatest factor in the emigration; by 1950, more than half of all Greek survivors had moved to Israel. While Jews linked to the Left were barred entry to Western democracies, all Greek Jews—no matter their political orientation—found that emigration to Israel was the only way to escape the draft and leave Greece legally.

Andreas Papandreou's government-in-exile (based first in Alexandria and then, at the end of the war, in Naples) returned to Greece after liberation to find the country largely under the control of the EAM. Immediately, it banned the EAM's military arm, Ethnikós Laïkós Apelefthero-tikós Strátos (ELAS), from Athens, and demanded that resistance groups disband and disarm by a December 10 deadline. EAM ministers withdrew from the government in protest. When permission for a leftist demonstration was revoked, the protesters went ahead and held it anyway, late on the morning of December 3, 1945. Government troops fired into the crowds, killing several participants and sparking what was to be a full-blown civil war, not concluded until 1949.

In short order, haggard survivors who had recently found themselves in Auschwitz and other camps were now confronted with the order to take up arms on behalf of the royalist government. Many survivors who had received a cold welcome home from their Christian compatriots commented bitterly on the irony of this turn of events. As Yitzhak Bileli, a Corfiote whose family was gassed in Birkenau put it, "I was sent to the camp as a Jew, and now for serving in the army I am Greek?"[92] On the other side, too, there were also ironies: many who had spent the war fighting in the resistance remained in the mountains, now to fight the Greek army in the Civil War. Solomon Matsas—who had fled the holding camp at Larissa, escaping deportation to Auschwitz—was among a number of Greek Jewish Communists who died in the Civil War. Those who survived it were targets of the postwar royalist reprisals. Some were executed or imprisoned in Greek detention centers; others were shunned by society, and found themselves without employment, friends, or social standing.[93] Into the 1950s, highly publicized trials of prominent leftist leaders furthered the view that the Greek Left was made up largely of pseudo-Greeks, traitors to the nation whose real loyalties lay with Moscow and other outside entities.[94] Escape from one fate led ineluctably to another.

Already within months of the liberation, leaflets were published documenting wartime atrocities perpetrated by the Greek Left during the occupation. Many suggested that the Left was no better than the Nazis, publishing photographs of mutilated and raped bodies as well as victims of torture. "It is your duty to expose the crimes herein recorded to all your friends and acquaintances," one text urged. "These are not photographs of victims of Dachau or other concentration camps of totalitarian Nazism. Who are the victims? Greeks. Who are the butchers? Totalitarian Greek Communists. . . . Draw your own conclusions. What is true of Dachau and Buchenwald is true of the KKE-EAM-ELAS."[95] As Greece became a key battle ground in the emerging cold war, and with millions of dollars of reconstruction aid pouring in from the West, the country fell under greater international scrutiny and control than at any time since independence. The politics of the Civil War combined with the acute outside interest in Greece's domestic affairs to create an extraordinarily conservative domestic political environment that bound together the monarchy, the Orthodox church, and the rightist political establishment more closely than ever before.[96] The institutions of punishment and rehabilitation of former leftists provide an example. Makronissos, the island detention camp where former Communists were imprisoned, held a number of Jews.[97] As part of the effort to reconstruct prisoners as good Greek citizens, an Orthodox seminary was planned for the island on the theory that becoming Christian was a vital component to "[re]becoming Greek."[98] As in the wake of independence, a sense of order was imposed by a national narrative that emphasized Orthodox Christianity as the core of the nation.

Obviously, the political climate was disastrous for anyone who had been active in the resistance or was a known Communist. But for those with no Communist ties whatsoever, the situation was also fraught. Many

Jews felt loyalty to the Soviet Union, for instance—not out of any political sentiment but rather just out of gratitude for the forces who had liberated a number of the Nazi camps, where they'd treated prisoners kindly, giving them food and warm clothing. Many of the Soviet soldiers were Jewish themselves and had shown particular kindness to the liberated Jews. Greek Jews who returned to Greece wearing donated Soviet coats, adorned with the hammer and sickle emblem, were imprisoned as Communists.[99] The pro-Soviet sentiments of Holocaust survivors were taken as a matter of politics, and set Greek Jews apart as suspect in the eyes of Christians. As one survivor recounts, he had tensions with Christians over everything, "even . . . because they were anti-Soviet, and I couldn't understand how they could feel this way towards those who had saved us."[100] In some places, largely those where anti-Semitic sentiment had been more prevalent, Greek Christians suspected that camp survivors had been "indoctrinated with Soviet ideology."[101] At the same time, politics created huge tensions within the Jewish community itself. Members of the resistance were regarded with contempt by Jewish survivors who had been deported to the camps or had gone into hiding with Christian families. International Zionist organizations were scathing in their criticism of the Kommunistikó Kómma Elládas, known as one of the most anti-Zionist Communist parties in the world.[102] The Palestinian Jewish press went so far as to single out Jewish Communists as the only real oppressors of Jews to be found in Greece.[103]

JEWISH OR GREEK?

The Greek Civil War dealt a serious blow to attempts to reconstruct Greek Jewry. Here a convergence of events was vital: the Greek draft, to which all men under the age of forty were subject; the foundation of the state of Israel in 1948; and the collusion of the Jewish Agency with the Greek government in passing legislation that would release Greek Jews from military service on the condition that they give up Greek citizenship and move to Palestine.

In 1945, many returning Greek Jews felt that they were coming home to their motherland. The relative indifference with which they were greeted was all the more painful for the loyalty they felt for Greece. The military draft was the final insult. Testimonials and memoirs tell the same story over and over again.

Gedalia Levy returned to Salonika on November 5, 1945. "No one took care of us. We were very disappointed, because we felt fully Greek [yavanim lechol davar]." Levy left for Palestine after being drafted into the Greek army.[104]

Isidor Alalouf, who survived the camps and returned to Salonika, was otherwise happy to be in Greece. But the prospect of military service was too much. "In 1949, I gave up my Greek citizenship just so I wouldn't have to serve in the reserves in Greece, and I went to Israel."[105]

Yaakov Jabari, a survivor from Didimoticho in the Bulgarian zone, initially settled down back in his hometown. But he, too, couldn't bear the idea of the draft, and closed down his business and left. "I went back [to Didimoticho]. . . . And together with Hayyim Kalbo from Nea Orestiada we opened a business selling Ouzo. We worked together until I had to enlist in the Greek army. At that time, the law that those who registered in Hakhshara are exempted from the draft was enacted, so I chose to go to the Hakhshara and not enlist, and then I went to Israel [in June 1946]."[106]

Alberto Gatenio returned from the camps to Salonika, where he got married. "I had the option of going to Israel, but my wife decided that it was better to stay in Greece. Within a few months, I found myself drafted into the Greek army as a regular soldier. Now I was involved in the Civil War in Greece, in an effort to stop the Communists who wanted to rise to power." For three years, Gatenio did not see his wife or newborn son.

> I was thrown into one of the worst regiments in the Peloponnese. It was a war of barbarism. We burned villages, exterminated, and looted. I felt myself as a Jew, and I wanted to get out of the army. In one of her letters, my wife wrote that Israel had been established, and that whoever wished to make aliyah had to give up their Greek citizenship. I decided to give up my Greek citizenship. I did it through the military bureaucracy, and I was released.

Gatenio met up with his wife and child, and moved to Israel in 1950. "I willingly gave up my Greek citizenship. Greece for me was a source of suffering. That is why I never went to visit and I never will."[107]

What these survivors all refer to was citizenship legislation that enabled Jews to avoid the draft by renouncing their Greek citizenship and leaving the country. Subsequent legislation granted a pardon to former partisans on the same conditions. Preliminary groundwork for agreements of this sort had been laid in conversations between the Greek prime minister and the chargé d'affaires of the Jewish Agency, Moshe Shertock, who had urged the Greek government to send all orphaned Jewish children to Palestine and argued that releasing Jews from military service would "assist voluntary emigration from Greece to Palestine."[108] During the course of the war, some Greek Jewish groups had tried to head off arrangements of this sort. In autumn 1944, for example, the Greek Jewish community of Cairo proposed a motion that denied the right of any world Jewish organization to speak on behalf of all Jews, affirmed that Greek Jews were first and foremost members of the Greek polity, and demanded that citizens be treated as individuals rather than as members of one or another religious collectivity. The proposal rejected any suggestion that the existence of a Jewish state should undercut the citizenship rights of Jews living in other states. "The existence of a Jewish home in Palestine [must] not have any adverse impact on the free exercise of the political and civil rights of persons of the Jewish religion resident in other countries."[109]

What was really at issue was this: Were Greek Jews more Jewish or more Greek? The question confronted Zionists and opponents alike. The pro-Zionist Greek community in Palestine, for instance, objected to the fact that the rights of its members as Greek Jews were being subverted by the Greek government, which in autumn 1944 had appointed boards of management for the Jewish communities of Athens and Salonika "without at least consulting the Jews who have fled from Greece to Palestine, who today number many thousands and are the largest cohesive section of the once-populous Jewish communities of Greece."[110] Both the Zionists and the anti-Zionists were thus in agreement that the fact of Zionism ought not one way or another to undercut their status and influence as Greeks. But both sides lost out. Partly because the circumstances of the Civil War upped the ante, the Greek government passed precisely the sort of legislation the Egyptian Greek Jews had lobbied against: Greek Jews could go to Palestine, but only if they ceased to be Greek.

Zionist agents urged Greek Jews to take this course, and set up camps (*mahanot Hakhshara*) around the country that prepared Jews for emigration to Palestine and taught them about kibbutz life. Representatives of the Magen David toured Greece and reported that while Greek authorities seemed to be doing everything they could to help survivors, it was nevertheless imperative that as many as possible immigrate to Palestine without delay. Jewish life in Greece was a wasteland, they said; there was no future for Jews in Greece.[111] Magen David Adom and Jewish Aid groups worked in association with UNRRA to foster emigration. The AJDC intervened with the Greek government for orphans to be sent to Palestine.[112] Ḥaluts (Pioneer, an organization that prepared people for the move) established a branch in Salonika in early summer 1945, and Ḥaluts and ha-Teḥiyah (Rebirth, a Zionist youth organization) soon opened branches in towns around the country. A Zionist library in Greek was established, and newspapers and pamphlets were distributed outlining the advantages of life in Palestine, and urging Greek Jewish survivors to migrate east.[113]

As more and more Greek Jews left for Palestine, the role of Greek Jewish agencies became less clear. KIS, for example, came under attack for being party to restitution plans that included provisions that revenues from unclaimed Jewish properties go to Israel. One plan, negotiated with the Israeli consulate, allowed for the transfer to Israel of up to 30 percent of the proceeds of heirless property.[114] Another was to establish special Greek bonds to underwrite Greek Jewish settlement in Israel.[115] Opponents called such plans "devious," and argued that Greek Jewish monies, such as they were, should be used to help the plight of Jews in Greece, not assist their transfer to Israel.[116] But the fact of the matter was that by 1950, more than half of all people who called themselves Greek Jews no longer lived in Greece.

Greece's resolutely anti-Israeli stance at the United Nations, a policy governed by Greece's concern over the status of the sizable Greek diaspora community in Egypt, heightened the problematic. When General Nikolaos Plastiras, a contender to lead the new Greek cabinet, said in an interview "that he look[ed] forward to the closest relations with Israel," averring that it was to become the center of the "revival of Near East civilization," Greek Jews hoped that they would benefit.[117] David Ben-Gurion, an avid independent scholar of classical Greek, paid a visit to Greece in late 1950, billing it as a private trip. Hopeful speculation, both among Israelis and Greek Jews, was that it might lead to a rapprochement.[118] Ultimately, though, Greece's Jews would become increasingly compromised by Greece's anti-Israeli stance and the growing assumption in certain Greek circles that all Jews were Zionists whose true loyalties lay with Israel.[119]

As had been the case many times in Greek Jewish history, those who survived trauma emerged from it only to find themselves trapped between various movements and factions, blamed by all sides for decisions they had taken earlier in the effort to save themselves. Leftists were condemned as both anti-Greek and anti-Zionist; Zionists were regarded by many Greek Christians as disloyal citizens; anti-Zionists were regarded as disloyal to the dreamed-of Jewish state. By now, these accusations were almost hackneyed, they had been leveled so many times before. And all this chaos was set against the backdrop of a devastated country and the raging Civil War. For Greek Jews, 1940–50 was experienced as a nightmare, a decade of bitter irony. First sent to the Albanian front, only to return home to the German occupation and deportation, those who survived the death camps came home to the Civil War or emigrated to Palestine to be drafted into another war. The long century of Greek expansion had concluded not with the chance for consolidation and normalization but with its opposite.

For those who had recently escaped war zones, barbed wire, and violence in Europe, what they found in Greece and Palestine was not as peaceful as they had hoped. Just as other major calamities in Greek Jewish history had been succeeded by others—World War I by the Balkan Wars, the Balkan Wars by the great fire, the great fire by the population ex-

Fig. 9.3. Salonikan Jews celebrate the declaration of independence of the state of Israel, 1948, Salonika, Greece. Permission Film and Photo Archive, Yad Vashem, Jerusalem.

changes and the heightened communal strife they brought with them—now the calamity of the Holocaust was followed by civil conflict and confusion almost everywhere the survivors turned.

Chapter 10

HELLENIZED AT LAST: GREEK JEWS

IN PALESTINE/ISRAEL

PEOPLE WHO HAD just suffered a trauma of literally unimaginable proportions had little time to heal. Those who stayed in Greece had to compete
with the acute needs of the general population, which had suffered immea-
surably during the occupation. And those who went to Palestine were
confronted with a chaotic and turbulent situation, a violent war between
Jews and Arabs, and a Jewish Ashkenazic elite that would regard the
Greek Jews with suspicion and as inferior.

In the midst of the chaos of postwar Greece—utter poverty, a bloody
civil war, and tremendous international pressure—its Jewish population
was reduced, almost literally, to a footnote. Embedded in subsection 4 of
a slim booklet titled *Greece, Basic Statistics: 1949*, is a terse paragraph:

> JEWS—Before the war, the Jews had 30 synagogues and 10 confes-
> sional schools. The status of the Jewish Communities in Greece is
> regulated by Act 2456. Each community has its Grand Rabbi. The
> number of Jews in Greece has been greatly reduced as the result of
> the deportations during the German and Bulgarian occupation. Only
> some 8,500 Jews have survived out of a pre-war total of 67,630.[1]

For most of these survivors, life in postwar Greece proved untenable,
and at least half left for Palestine. But once in Palestine, they left in
record numbers; Zionist organizations there painted the Greeks as
"lax," "unpatriotic," and "frightened of hardship"—a particularly
ironic claim, given the inexpressible hardships they had so recently endured. The story of Greek Jews beyond Greece, especially in Palestine/
Israel and more recently in the Israeli cultural imagination, is the final
chapter in the complicated story of Greek Jews postwar. It is also the
final chapter in the Hellenization of Jews of Greek descent. In Auschwitz,
Jews from Greece—Romaniote and Sephardic alike—were uncomplicatedly Greek. In Palestine/Israel, the next site of mass heterogeneous Jewish habitation, the nationalized Jewish categories of the war were retained and developed. As in Auschwitz, no "hyphenated identities"
existed. There were simply Greeks.

PALESTINE/ISRAEL

Culturally, Palestine/Israel was an alien environment for the Greek Jews who went there. Between 1933 and 1940, a total of 280,000 Jews had fled from Europe to Palestine (in comparison to 155,000 to the United States, and about 100,000 to Mexico, Central America, and South America combined). Less than 1 percent of those who went to Palestine in the pre- and early war period were from Greece. Little money was set aside for the Greeks: in 1940, a total of almost £50,000 was spent on immigrant needs; £331 of it was designated for Greeks, together with Yemenites.[2] After the war, when immigration surged, the numbers of Greek immigrants remained tiny in comparison to other national groups. In 1948, 192 Jews came to Israel from Greece; in 1949, 1,364 more arrived. The next year added another 463. Of the total number of Jews who immigrated to Israel between 1948 and 1950, only .04 percent was from Greece.[3] In subsequent years, the numbers of Greek Jews coming to Israel were proportionally so tiny that the annual reports of the Israel Office of Information grouped them in the general category "other countries." In the first year and a half of independence, Israel's population increased by 50 percent. "One out of three Israelis was therefore a newcomer and a stranger. They were also strangers to each other, having come from almost every country in the world and speaking any number of different languages."[4] For Greeks, who came in small numbers, the sense of isolation was only more acute.

In Greece, debates between Zionist and anti-Zionist organizations over immigration to Palestine were particularly fierce. The Jewish Agency and the Greek consular authority were in official opposition to the resettlement of Greek Jews, and UNRRA in Greece was criticized for "overstep[ping] the framework of its duties . . . [by] encourag[ing] the departure of the Greek Jews." (Meanwhile, UNRRA in Palestine was accused of anti-Zionism, and its agents suspected of "influencing [Greek] refugees to return to Greece by promising them every possible facility," and helping with visas and travel arrangements. A column in the Palestinian Jewish newspaper *Davar* asked, "Has UNRRA deteriorated into a mere travel agency?")[5]

An array of critics, Jewish and non-Jewish, condemned Zionist organizations for fostering immigration without "tak[ing] steps to insure that [immigrants] . . . were provided with opportunities to live comfortably and make a living." By late 1945, a number of Jews who had come to Palestine expressed the desire to return to their countries of origin, further compromising the Zionists' position. As the Greek consul for Palestine and Trans-Jordan put it, "The princip[al] argument for the formation of a Jewish state in Palestine put forward by the Zionists—that is, that the Jews are, so it is claimed, unable to continue living in Europe since life there has become intolerable for them everywhere—is *ipso facto* demolished when the persons concerned acknowledge in practice that they prefer to return to their homelands in Europe rather than continue to reside here." Greek Jews who were hesitant about moving to Palestine—or worse, who wished to return to Greece once they got there—were placed in an all-too-familiar situation, accused as "traitors." Whereas earlier they'd been Greek traitors, now they were Jewish ones. "For the extreme nationalist and terrorist Jewish organisations, any Jew who makes his way to Palestine and later leaves the country is a traitor to the Jewish ideals."[6] The double-edged dynamic that had prevailed in Greece continued in Palestine, simply in inverted terms.

Palestinian Jewish newspapers published articles about the particularly high numbers of Greek Jews in this situation. In Palestine they couldn't find work, had no housing, and didn't know the language. Many held property in Greece and wanted to go back to claim it. *Ha-Aretz* condemned them, saying that "these people ought to have learned a good lesson from their tragic past, and not attempt to return to Greece."[7] But quite aside from things in Greece that called the recent émigrés back, there was plenty about the circumstances in Palestine to drive them away. Already by 1937, fifty thousand guns had been smuggled into Palestine, along with hundreds of thousands of rounds of ammunition. Violence between Arabs and Jews was a daily occurrence.[8] In Tel Aviv, crime had become so frequent that the city was dubbed "the Chicago of Palestine."[9] The Arab rebellion brought numerous bombings between the two communities. At the start of World War II, the region was already in a "state of political, economic and psychological unrest." Press reports lamented that "encounters, sabotage, kidnapping, bomb explosions, derailment of trains and trolleys, sniping and armed robberies have become daily phenomena."[10]

A decade later, when Greek immigration rose, the situation was more volatile still. By 1947, portions of the country felt like a prison. "Jerusalem is criss-crossed with barbed wire. Large areas can be entered only by people equipped with special passes, and the ordinary public has to make its way about the town as best it can." The Jewish press complained that in dividing up the city in this way, the British had created a situation in which "the possession of one's home and the continuation of one's family life depend on . . . basically meaningless distinctions [between different nationalities]."[11] A dismayed Greek observer, Eliahu Zacutta, commented ironically that "barbed wire . . . [had] become so typical of Palestine as to justify its becoming its emblem." In an attempt to contain the situation, the British placed Tel Aviv and Jerusalem under martial law, fenced off heavily populated areas, stopped mail delivery, and prohibited incoming and outgoing phone calls.[12] Curfews became a frequent occurrence.[13] The violence was not simply between Jews and Arabs. Zionists battled with anti-Zionists, and religious Jews fought with the secular political establishment. In spring 1951 Brit Kannaim, an ultraorthodox organization, plotted to bomb the Israeli Knesset.[14] Other ultraorthodox and terrorist groups, such as the Stern Gang and Naturei Karta, also agitated against the government.

Things were also difficult economically. For decades, Palestine had suffered from high unemployment rates, especially among new immigrants. While Salonikans who had moved there in the nineteenth century flourished, leaving their mark on various professions, newcomers found work and housing difficult to come by.[15] In the mid-1930s, hundreds of people, Jews among them, had applied to the Italian consulate in Jerusalem in the hopes of securing employment in Ethiopia.[16] In the early 1930s, tourism had been Palestine's most important source of income after the orange industry; tourism had been demolished by the decade's end by fears about Arab-Jewish terrorism.[17] In the wake of World War II, the situation worsened, largely because of the massive immigration of the postwar period. The population of Israel doubled between 1948 and 1952. The extraordinarily complicated and difficult circumstances under which this occurred have been amply documented, most infamously in Tom Segev's *1949: The First Israelis.* "The immigrant becomes a number," wrote one journalist, an anonymous speck in a huge crowd that lived in the squalor of the transition camps for months on end.[18] The price of the 1948 war also continued to be felt.[19] The huge influx put immense pressure on the economy at the same time that earlier sources of income, notably tourism, all but dried up.

The circumstances were made more difficult by the fact that what was termed Palestine's "absorptive capacity"—its ability to take in immigrants—was contingent on settling large swaths of barren territory, particularly the Negev desert region in the south. While immigrants in the 1930s had largely moved to Tel Aviv and Haifa, by the 1940s more and more were sent to rural areas. They worked cultivating citrus, avocados, and bananas, running fisheries, and establishing agricultural settlements in remote regions.[20] Dominant groups of immigrants, those who had come to Israel before the war and had the force of numbers, founded their own neighborhoods, such as the Romanian Settlers' Association in Tel Aviv.[21] Greek and other groups that came later—almost all Sephardic—had a more difficult time finding housing.

Even for the many Greeks who were happy to be in Palestine, things weren't easy. Immigrants who arrived illegally on ships were seized by the

British royal navy; those who were found after they had touched ground were herded on to deportation boats and sent away. The most famous of these illegal boats, of course, was the "Exodus 1947," which had over forty-three hundred central European Jews on board. Widely circulated photos of the ship teeming with refugees and forbidden from docking were a major catalyst for the creation of Israel in 1948. While Zionist organizations enthusiastically propagandized in Athens, Corfu, Patras, and Salonika, many of the Greeks who responded by packing up for Palestine did not know that they might not even be allowed in. *Haganah* ships were turned away in Haifa as the United Nations Special Committee on Palestine representatives looked on; illegal immigrants were sent to camps on Cyprus or back to displaced persons camps in Germany.[22]

Yet many of the Greek immigrants, like hundreds of thousands of other Jews, felt that Palestine/Israel provided their only chance at some sort of security. They also found its Mediterranean climate comfortingly familiar.[23] While culturally alien, geographically it offered familiar features of home—proximity to the sea, easy winters, and mountains—blended with something new: perceived security for Jews. Centuries earlier, Salonika had presented Iberian refugees with a similar sense of at once familiar and alien refuge. Now yet another move, to an outpost still further east in the Mediterranean, conjured up many of the same feelings. Pavlos Simcha's family left Kavála in 1940, when his father decided that he needed to find a better place for his children to live. His family was one of the few to realize so early that Bulgaria's alliance with Hitler, combined with its proximity to Thrace, would bring horrors to the Jews of the region. Many years later, in 1988, Simcha likened prewar Thrace to Israel, a place where Jews lived happily, but were surrounded by enemies on all sides. Israel reminded him of home. Even though Kavála had been a wonderful place to live, Simcha recalled, security in Israel was much better.[24]

Emigration from Greece to Israel continues to this day, largely because of the lack of marriageable Jews in Greece. In 1958 Esther Pitson, one of the few Jews to resettle on Corfu after the war, had one daughter married to an Israeli and living in Israel, and was planning to send her second daughter there to find a husband as well.[25] Loutsiana Modiano-Soulam, who was sent to Auschwitz from Rhodes in the last months of the war, returned to Rhodes, where her husband served as *shamash* in the synagogue. In an interview in 1968, she concurred that she too had as her "dream to send my three daughters to Israel, because here they won't find a husband and make a family."[26] Many of those who have stayed in Greece have seen their children marry Christians, although not all are sorry about it. Maty Azaria, of Salonika, returned from the camps to resettle in Greece. Both her daughter and her niece married Christians in civil ceremonies. "Both of them got very good guys. They love us, they respect us."[27] But for those who don't like the idea of intermarriage, Israel has long seemed like the best option. And while there are relatively few Greek Jews in Israel, the Greek as a cultural type has taken on large dimensions in the modern-day Israeli cultural imaginary.

"Singing Greek": Greek Jews in the Israeli Cultural Imagination

> I feel like an idiot—I came here from Salonika thirty
> years ago, and I have nothing! You [Ashkenazim] show
> up, and you get everything right away.
>   —Salamonico to Tadeusz Berezovsky, in the
>   Israeli film *Salamonico*, 1972

In the 1972 Israeli film hit *Salamonico*, a kindhearted but boorish Greek Jew tries to make his way in mainstream Israeli society.[28] At the insistence of his social-climber wife, Salamonico moves his family out of the port slums to a fashionable north Tel Aviv neighborhood, where his new Ashkenazic neighbors scorn and revile him. His housewarming party, which draws scores of Sephardic friends and relatives and features live bouzouki music, keeps the whole building up all night. The neighbors complain that they can't figure out who is and who isn't a member of his family— "there are so many of you." While Salamonico insists on keeping tradition, his children rush off to parties on Friday nights, interrupt him at the dinner table, and ignore his wishes. When his teenage daughter is taken advantage of at a party by an Ashkenazic boy and becomes pregnant, Salamonico demands that they marry. The boy's parents are shocked by the suggestion, telling him by way of counterproposal, "We'll be happy to pay for an abortion." Salamonico prevails, and in a (far better) forerunner to *My Big Fat Greek Wedding* the event brings together Salamonico's huge, fun-loving family and the boy's uptight European relatives, who gawk as the Greeks dance the *tsifteléli*, down gallons of ouzo, and generally show that they know how to have a good time.

Salamonico is the Israeli version of the quintessential Greek. He works in the port, and has a huge mustache, large family, simple mind, and big heart. He is loutish and Oriental; when a Polish friend takes him to a French restaurant, Salamonico causes a scene. Disgusted with the food and wine, he calls for arak and meze in their place, and is reprimanded by the waiter, "We are civilized here." He's lived in Israel for decades, but remains cocooned in his Greek world, hanging out in tavernas, fraternizing with fellow Greeks, and never learning to read. His speech is peppered with Greek, Ladino, and Turkish expressions, and he and his wife are preoccupied with the Greek royal family, whom they regard as the pinnacle of class and elegance.

The picture is a peculiar composite of prewar Jewish Greece and 1970s' Israeli visions of the country. Salamonico is also a pastiche of Sephardic and Romaniote elements; the tensions between the "foreign" aspects of Sephardim vis-à-vis the "more Greek" aspects of Romaniotes that had earlier played a strong role in Greek Jewish culture are fully elided in the person of Salamonico. Features of Salamonico's character as cast by the Ashkenazic filmmakers who created him are quintessential traditional Salonika: his language, profession, and attachment to custom.[29] But others are pure 1970s' working-class Greece. The Greek folk/pop music and Zorba-style dancing at his parties, Salamonico's physical appearance (complete with Greek fisherman's hat permanently affixed to his head), and the food and drink—on frequent display throughout the film—are all straight off a tourist brochure. Key features of traditional Salonika are ignored: the cosmopolitan and educated nature of many of its residents, its intense involvement with its surrounding environment, and above all its community's sense of superiority to other forms of Judaism. In *Salamonico*, the Greek Jew has been simultaneously Orientalized and Hellenized.

Despite their small numbers, Greek Jews have come to occupy a large space within the Israeli cultural imagination. The image of the Salonikan stevedore—the good-hearted, strong, and simple man of the sea—was early on coined as a cultural archetype, although the term Greek was not used to describe Salonikan immigrants until after World War II.[30] The Salonikans were long recognized as superior commercial fishers, the first Palestinian Jewish commercial fishers learning the trade mainly from them as well as from Italians, Romanians, and Arabs.[31] For decades, movies, books, and comedy routines have referenced the Greek of the docks. The actor who plays the role of Salamonico—Reuven Bar-Yotam—is Moroccan, and built his career largely on a comedy routine in which he played a good-hearted Greek. The Greek Jew, at once Sephardic and European, has proven a comfortable symbolic bridge between the often-alienated worlds of central and east European Jewry, on the one hand, and on the other, the heavily Arab culture of the more than one million *mizrahim*, or "Oriental Jews" who were brought to Israel from Yemen, the Maghreb, Egypt, Iraq, and other Arab regions in the early 1950s. The Greek Jew again mediates between East and West.

Israel's relationship to its Middle Eastern setting is an ambivalent one. Even as Israeli culture glorifies and appropriates its more benign trappings (for instance, falafel, an Arab dish, has been declared the national food; the Palestinian cactus is depicted as a metaphor for the rugged Israeli character; and the Egyptian singer Um Kulthum was recently the pinnacle of cool), the more problematic features of its Arabness are reviled or ignored. Just as in France in the wake of its colonization of North Africa the material trappings of Maghrebi culture rose to the highest in vogue, mainstream Israeli culture has adopted and idealized various Arabicisms while distancing the Arab. Many of these Arab cultural adaptations have been mediated by the mizraḥim, whose own culture is largely an Arab one. Yet mizraḥim have long been perceived as inferior outsiders in Israel's decidedly Ashkenazic, Western culture; behind the high-profile conflict of Israelis and Palestinians lurks a societal fracture between European and "Oriental" Jews.[32] The Israeli cultural embrace of idealized Greekness can be read as an unspoken attempt to bridge the gap. Greeks, put simply, have often functioned in the Israeli cultural imagination as "safe" mizraḥim.

Among other things, this has spawned the peculiar Israeli phenomenon of the fake Greek singer, best typified by Zohar Argov, the hugely popular Yemenite performer of the 1970s and 1980s. Zohar Orkabi was born in Rishon LeZion, Israel, in 1955. Like many other mizraḥim, Argov changed his obviously "Oriental" family name to a benignly Israelified and Ashkenazified one. In the 1970s he rose to fame with the song "Elinor." A Hebrew version of Stelios Kazantzidis's Greek single "*Ipárho*," "Elinor" is probably the biggest mizraḥi hit ever. Argov also recorded countless other Greek songs, some in Hebrew and others in a garbled and scarcely intelligible Greek; in 1984 he starred in the musical film *Kasach*.[33] By many ardent fans, Argov is regarded as a martyr to the ills of Ashkenazic antimizraḥi sentiment. While he attained success, he was imprisoned for drug possession and committed suicide in prison in 1987. One of many Web sites dedicated to his memory features as its background music Elton John's "Candle in the Wind"; written in homage to Marilyn Monroe, the song's lyrics—"They put you on a pedestal, and they made you change your name"—invoke the image of the appropriated, exploited, and misunderstood superstar whose success depends on clever concealment of humble origins.[34] Like Monroe, Argov's greatest fame came after, and in part because of, his early, self-inflicted death—a death that reflected the inner torment brought by the effort to escape his roots. A 1993 documentary on Argov, *Zohar: Mediterranean Blues*, teases out these same themes.[35] The songs Zohar sang were traditionally Greek—not Greek Jewish or Sephardic but the popular hits of contemporary Greek culture. Scores of other Israeli "Greek singers"—some of Greek origin, but most, like Argov, mizraḥim from Arab countries—have championed this music as well.

In the late 1950s Aris San, a Greek Orthodox Christian singer from Salonika, began performing in Haifa after deserting the Greek military and fleeing Greece. San later became a mainstay of a Salonikan Jewish-owned nightclub in Jaffa, the Arianna, known for featuring Greek music. His arrival on the scene coincided with the influx to Israel of hundreds of thousands of mizraḥim. At the time, "Greek laïka [pop/folk] music provided a legitimate way to publicly enjoy the type of sounds beloved by Jews from Arab countries."[36] San (who appears in *Salamonico* as the entertainment at Salamonico's housewarming party) and other Greek performers soon inspired mizraḥi musicians to perform their own "Greek" music. Argov, Stalos, Nikoles, Haim Moshe, and Levitros—all built their careers singing Greek hits and built a Greek identity around their original, generally mizraḥi selves. Levitros, for example (whose album jacket announces his music to be "Oriental"), is Israeli born, of Iraqi origin. He Hellenized his original name, Levi Mu'alem, by adding the Greek-sounding suffix "tros," and explains that he "was not born Greek" but became Greek through Greek music, which "penetrates to every place and every person."[37]

Greek music provided Argov and other mizraḥi singers with an "acceptable" outlet for the expression of their Oriental origins. The Greek language, which few of them knew, substituted for Arabic—which for many was a second, if not first, tongue. By "becoming Greek," they toned down their Arabness and moved themselves further toward the West on the Oriental-European spectrum. Overwhelmingly, their audience was mizraḥi. Just as the classification of all non-European Jews in Israel as Sephardim (despite the fact that most mizraḥim have no ties to Spain or Ladino culture) has sanitized and minimized the markedly Arab features of their culture, the embrace of Greek music was a means through which mizraḥim found a nonthreatening, semi-Europeanized cultural mode of expression.

At the same time, and for different reasons, Ashkenazim, too, have come to embrace Greek music. In the 1980s the rock singer Yehuda Poliker, a mainstay of Ashkenazic top-ten lists, "recovered" his Greek origins. Poliker's family, originally from Salonika, was part of Israel's gritty and largely *sefardi* working class, but his grunge-style music reflects the tastes of the Ashkenazic mainstream. In his thirties Poliker became intensely interested in his Salonikan roots, publicly tracing the multigenerational impact that the Holocaust had had on his family. Increasingly, the central themes in his music were the Holocaust and the Greek Left; Poliker translated and performed several partisan songs. In exploring his past and his Greekness, he argued that Israel needed to embrace its identity as a Mediterranean country, while his resurrection of the Greek Left served as a commentary on contemporary Israeli policies, particularly toward Palestinians.[38] As part of his personal transformation, Poliker developed an interest in contemporary Greece, bringing famous singers like Glykeria

Fig. 10.1. Back cover of Levitros's *Greatest Hits, Vol. II*. Permission Azulai Brothers Productions, Israel.

from Greece to Israel to perform with him. In Poliker's case, Greekness—an authentic Greekness, as Poliker pointedly emphasized—became a means of complicating the category Ashkenazic. While Oriental singers Europeanized themselves via Greek music, later performers like Poliker saw Greek music as a vehicle for positive self-Orientalization and a means of interrogating the dominant categories of Israeli culture.

These two embraces of Greek music have more recently moved closer together, spawning a new wave of "Greek" singers, among them Shlomi Saranga (*Greek Celebration*), Yehuda Saleas (*Best Hits from Greece*), and Boaz Tabib (*Crazy Greek*). Greek standards like "*Fíge Fíge*," "*Óhi Óhi*," "*Min Periméneis Piá*," and "*Páploma*" are well-known in Israel. Descendants of Greek immigrants, too, have more lately come on to the scene; Stelios Kanios, the child of Greek immigrants, has covered countless Stelios

Kazantzidis hits, and his albums have sold tens of thousands of copies. Kanios prides himself on his Greek "authenticity" and the fact that he sings in Greek.[39] The genre into which they fit is now known as "Mediterranean Music" (*musika yam-tikhonit*), a compromise between East and West, described by a geographic term (Mediterranean) that doesn't carry the more loaded connotations of Ashkenazic or Oriental.[40] A few "real" Greek singers (that is, non-Jewish Greek singers from Greece)—notably Glykeria, Georgios Dalaras, and Haris Alexiou—are popular, particularly among Ashkenazim, but for the most part the Israeli Greek music market is dominated by the Israeli version. (And one of Greece's most famous musicians, Mikis Theodorakis, is persona non grata in Israel, where the press frequently rails against his ardent support of the Palestinian cause and his provocative statements about "world Jewry.")[41]

Israeli-Greek music (of Argov's sort, and only to a degree Poliker's, which also draws from Greek laïka but is classified as "ethnic" rock) is a genre within *musica* mizrahit (Oriental music), a musical culture that "combines certain stereotypical musical elements from both Western and Eastern traditions," and tries "to appeal to a Western audience by developing a stereotypical Oriental sound that is pleasing to the Western ear."[42] Just as musica mizrahit as a whole tries to bridge the worlds of East and West, Greek music in particular has occupied an important place for Ashkenazim and mizrahim alike as a mediating, European-yet-not-European cultural form.[43] As David Margaritis, the manager of Radio Piraeus, an Israeli Greek music station broadcasting from Gush Dan (88.8 FM), says, "The secret of Greek music [is] . . . [e]veryone can plug into it—Palestinians, Poles, Iraqis, everyone."[44]

THE HOLOCAUST IN GREEK ISRAELI MEMORY

The mediating functions played by Greek music in Israel are an apt metaphor for Greek Jewish history as a whole, the status of Greek Jews in Israel, and the geographic location of Greece itself. The only major non-Ashkenazic group to perish in the Holocaust and one of the only major Jewish groups from Europe to be classified as Sephardic, Greek Jews in Israel exist in an ill-defined space between East and West that was frequently omitted from the Holocaust narrative. In the 1980s and 1990s, as the last survivors of the Holocaust moved into old age and began to die, the project of capturing as many testimonies as possible took on a new urgency.

In the mid-1980s *Lo Nishkah!* (We Shall Not Forget!) began publication. Issued annually in Tel Aviv on the eve of Holocaust Commemoration Day, *Lo Nishkah!* is the work of the Organization of Greek Survivors of

Extermination Camps in Israel, Division of Continuing Generations
(ארגון ניצולי מחנות ההשמדה יוצאי יוון בישראל: חטיבת דור ההמשך). Dedicated to preserving the memory of the Holocaust and Greek Jewish culture, *Lo Nishkah!*
gives us a view of how Greekness, Greece, and the arrival of Greek Jews in
Israel are remembered by second-generation Greek Israelis. A few themes
emerge. For one, Greek Jews in Israel recall the Greek aliyah as having
been far smoother than documents from the 1940s and 1950s indicate.
In the 1991 issue Shmuel Refael, the son of a survivor, wrote, "The Holocaust survivors born in Greece were absorbed quickly in Israel, while comfortably acclimating without burdening the establishment."[45] Another recurrent theme is the ignorance of most other Israelis regarding the
Holocaust in Greece. In the 1990 issue David Koen, son of Leon Koen,
commented,

> I was always asked, "What? The Jews of Greece were also in the
> Holocaust?" There weren't many articles or stories about the Jews
> of Greece. The media always talk about the Warsaw ghetto rebellion,
> the culture of the Jews of Germany and Poland. Weren't the Jews of
> Greece kosher Jews? According to my father's stories, the Jews of
> Greece were righteous, honest, [and] helped others.[46]

The exclusion of Greek Jews from the Holocaust narrative was linked
to their liminal status between the dominant categories—mizrahi/Ashke-
nazic—of Israeli society. Many of the testimonies gathered from Greek
Jews in the 1980s came from survivors who had lived in Israel for decades
by the time they were interviewed. Several filtered their recollections
through the sense of cultural liminality they had experienced in Israel.
Mano Avraham Ben-Yaakov of Salonika was one of many survivors interviewed in the mid-1980s. He closed his testimony—a grueling account of
life in Auschwitz—with a remarkable statement:

> Now, having revealed the goings-on and my history in the camps, I'd
> like to make some concluding remarks, that in my opinion deserve
> to be longer than [those about] the events themselves. We were led
> like all of Europe's Jewry to the crematoriums, and our fate was no
> different than the fate of all Jewry. However, in my opinion, Greece's
> Jewry was hurt more than any other community in Europe. At the
> time of liberation, no one imagined that there was any difference
> between the different [Jewish] communities. But despite all that we'd
> been through, after liberation we were met with a humiliating atti-
> tude. I refer here to the attitude of the Ashkenazic survivors, who
> withheld help from us when organizing services for the high holidays
> of 1945. The Ashkenazic survivors claimed that we were Reform
> Jews, and this was a terrible humiliation. And in my humble opinion,

it was already then that the breech between Sephardic and Ashkenazic Jewry was forged. But the Greeks do not give up.[47]

Ben-Yaakov identifies a dilemma that many Greek Jews had first encountered in Nazi camps and later in the displaced persons camps—a dilemma that in Israel was to become one of the defining features of their identity. The divide between Sephardim and Ashkenazim remains to this day perhaps the most striking, if unacknowledged, feature of Israeli culture. But while other categories of Jews—"Oriental" Jews from Arab countries, on the one hand, and central and east European Askenazim, on the other—could be readily identified and described, Greek Jews did not fit easily into the demographic and cultural schema. In Israel, Greek Jews had difficulty negotiating an identity that captured both their "true" Sephardic culture and their status as survivors of the Holocaust.

The absolute centrality of the Holocaust to this divide emerges clearly from the transcripts of the Adolf Eichmann trial, held in Israel in spring 1961. The trial proceeded according to geography: testimonies and evidence were brought in a sequence that followed the spread of the Holocaust, moving in a roughly eastward direction. Testimony from Greece came toward the end, or easternmost, point of the trial, following testimonies from Yugoslavia and Bulgaria, and preceding Romania and Slovakia. This geographic approach underscored the position of Greece's Jews between East and West. It also furthered their nationalization; the Central Jewish Council of Greece, which put forth the initial Greek testimony for session 47, which dealt with Yugoslavia, Bulgaria, and Greece, was "actually an official council which act[ed] in the name of the Kingdom of Greece," and its findings were treated "as an official report issued on behalf of the government."[48] Thus, for the first time since the end of Ottoman rule, the Jewish council was recognized as an official branch of the Greek government, albeit acting on behalf of a Greek Jewish population that for the most part had ceased some decades earlier to reside in Greece.

The council's document, no. 832, gave a numerical summary of the Jewish populations of Thrace, Macedonia, Thessaly, central Greece, the Peloponnese, Epirus, and the Greek Islands just before German occupation. It estimated that seventy-seven thousand Jews were in Greece before the deportations, of which ten thousand survived. The basis for the report was the diary of Yomtov Yekuel, legal adviser to the Salonikan Jewish community during the occupation. Yekuel was arrested by the Germans literally in midsentence, deported, and killed in Auschwitz. The diary itself, which contained lengthy commentary on Max Merten, Dieter Wisliceny, Alois Brunner, and other leaders of the Greek occupation apparatus, was also admitted as evidence. Composed in Greek, Hebrew and German translations of the diary were made for the court.[49] Testimony on the Holocaust in Greece was provided by a single witness, Itzchak Nechama,

Fig. 10.2. Spectators at the trial of Adolf Eichmann. Permission Film and Photo Archive, Yad Vashem, Jerusalem

who spoke only of his hometown, Salonika.[50] One result was that in the course of the trial, the identification of Salonikan with Greek became more complete.

Hanna Yablonka has recently written of the careful selection of witnesses at the Eichmann trial, designed to "shape the way in which the Holocaust was remembered through this highly publicized [event]." Witnesses "who came from especially interesting locales" as well as those who were particularly well educated and articulate were the top picks.[51] Nechama met both of these key criteria. Salonikans tended to be well-educated polyglots, and their status as Sephardim gave them an exotic edge. In addition to Nechama's oral testimony in the court, the Ghetto Fighters' Museum provided nine testimonies of Greek Jews living outside of Israel. In Israel, the Association of Greek Jews offered other names and possible witnesses.

The Eichmann trial came at the end of a decade fraught with conflict between mizraḥim and Ashkenazim. In addition to the greater global significance of the trial, the Israeli government viewed it as a way to educate mizraḥim about the Holocaust, "exposing them to a part of Jewish history with which they had had little or no contact."[52] In the words of the director of the prime minister's office, Teddy Kollek, "[T]he trial had another important function that we are not capable of appreciating. After all, a generation has grown up in Israel that did not know Hitler, and in our midst there are hundreds of thousands of immigrants from lands distant from Europe, who had no experience of the suffering about which we are hearing."[53]

While the Holocaust was an absolutely foundational feature of the modern Ashkenazic experience—and of Israel's Ashkenazic leadership—the more than one million mizraḥi Jews who had come to Israel in the 1950s had little knowledge of or interest in it. The division between Ashkenazic and mizraḥi responses to the trial echoed the divide between the two groups in Israeli society at large. While some mizraḥim did, as Kollek and others hoped, feel closer to the Ashkenazim as a result of what the Eichmann trial taught them, others wound up feeling once again marginalized and excluded from the Ashkenazic mainstream.[54]

For Greeks, who occupied a peculiar in-between status in this schema, the dynamic was especially confusing. Nechama, as the sole Sephardic witness at the trial, was to "embod[y] the oneness of Jewish destiny," his personal experience as a Sephardic Holocaust survivor binding together the destinies of Sephardim and Ashkenazim alike.[55] This lofty goal overshadowed the particularity of the Holocaust as experienced by Greek Jews, and subsumed the category of Greek Jewry under the broader rubric of Sephardim. The heads of the Israel-Greece league—Jean Allalouf, Maurice Ayyash, and Yitzhak Ben-Rubi—wrote an open letter to the editor of *ha-Tzofe* (Spectator) complaining that only one court session in the Eichmann trial had been given over to Greek Jews.[56] At the same time, non-Sephardic Greeks resented the fact that Greek Jew had become synonymous with Salonikan. Just as the Greeks as a whole felt written out of the Ashkenazi-dominated Israeli Holocaust narrative, Romaniotes (indigenous, Greek-speaking Jews, who were not Sephardic) felt excluded from the narrative of Greek Jewry.[57]

In multiple ways, versions of the liminality that Greek Jews had experienced in the Greek Christian context were replicated and reworked in the postwar Israeli one. In Greece, Romaniotes had struggled to differentiate themselves from Sephardim. In Israel, Romaniotes and Sephardim—now Greek Jews, a nationalized category—struggled to differentiate themselves collectively from other, non-European Sephardim *and* lay claim to their status as survivors of the attempted destruction of European Jewry. In Greece, Jews had felt more Greek or less Greek according to their cultural practices, the language they spoke, and whether they lived in Old or New Greece. In Israel, all were grouped as Greek Jews, even as the Greece they had left became, in the 1950s and 1960s, one of the most ethnically homogeneous and uniformly Christian nations in Europe. At the same time, in Israel the image of the Greek Jew became more Greek and less Jewish. The Salonikan traded his fez for a Greek sailor hat, danced traditional Greek dances, played the bouzouki, and drank ouzo. In Israel, he was more Greek than he'd ever been in Greece.

Chapter 11

CONCLUSION: GREEK JEWISH HISTORY—

GREEK OR JEWISH?

IN GREECE ITSELF, the Greek Jew as a category has in past decades slowly found greater purchase. The five thousand Jews living in Greece today are active preserving and documenting their history, and advocating for Jewish rights in Greece. The Jewish community of Thessaloniki, the Jewish Museum of Thessaloniki, and the Jewish Museum of Greece, in Athens, all issue regular publications, and the efforts of Greek Jewish groups within and outside Greece have recently led the Greek government to acknowledge the contributions of Greek Jews to Greek history, and to commemorate the devastating losses the community suffered in the Holocaust. In October 2003, a monument in honor of Greek Jewish soldiers killed in the 1940–41 war was established beside Salonika's Jewish cemetery. In May of that year, a plaque was erected at the entrance to Salonika's French Institute, which lost many students and alumni in the Holocaust. Greece for the first time ever observed a Holocaust Remembrance Day on January 27, 2004, the anniversary of the liberation of Auschwitz-Birkenau.[1] After intense negotiations, the German government has begun distributing pensions to Greek-born Nazi victims; under a previous agreement between the Greek and German governments, Greek Jews had been excluded from the restitutions paid to other survivors. Public figures have begun taking steps toward acknowledging the Greekness of Greece's Jews. In a Holocaust Remembrance Day speech, Foreign Minister Georgios Papandreou reminded his listeners that "in our country, Greek citizens, who also happened to be Jews, were arrested, loaded onto trains and taken to concentration camps that are today a symbol of the worst atrocities committed by humankind."[2] Similar events in 2006 emphasized the unity of Greek destiny and the need to see Greekness as a category that transcends religious distinctions.[3]

At the same time, anti-Semitism in Greece is by no means a thing of the past; in some respects, it is stronger than ever. And to this day it coalesces around the argument that Greek Jews somehow aren't really Greek. While in comparison to other European nations acts of anti-Semitic violence in Greece are relatively few, this says more about the persistent strength of European anti-Jewish sentiment than about the absence of it in Greece, where no sooner are commemorative monuments erected than they are defaced. In Athens, the Jewish Holocaust Memorial has been repeatedly spray painted with swastikas and the marble plaque smashed. A plaque marking the Square of the Jewish Martyrs, opposite the Theseion train station from whence Jews were sent en masse to Auschwitz, has suffered the same fate; in 2000, in a particularly heinous episode, close to a hundred graves in Athens' Jewish cemetery were spray painted with Nazi symbols and scrawled with graffiti that read, among other things, "Hitler was right." Synagogues, Jewish-owned businesses, and memorials in Volos, Jannina, and Salonika have been attacked and vandalized.

While the official church position regarding Judaism has always condemned anti-Semitism and has placed emphasis on Judaism's vital contributions to Christianity, over the years many church officials have published or publicly spoken on strongly anti-Semitic themes. This anti-Semitism often takes the form of a conflation of Zionism with Judaism, accusing Greek Jews of being anti-Greek or dangerous to the Greek state because of the supposed support of all Jews for the Zionist cause. In its most hysterical and hyperbolic form, such rhetoric syllogistically converts all Jews into Zionists; with dominant Greek public sentiment being strongly pro-Palestinian and anti-Israeli, Greek anti-Semitism often masquerades as a principled political stance. At the same time, the formulation discredits legitimate political positions—that Greece's stance on the Palestinian problem has been distinctly anti-Israeli is certainly not evidence that Greece is in some fundamental way anti-Semitic. Yet nor, of course, is the fact that some Greeks are Jews any sort of evidence that they are less than fully loyal to Greece.

Not surprisingly, the routine equation of Zionism and Judaism has led to outbreaks of anti-Jewish propaganda and activity that correspond to political events in the Middle East. Following the Israeli invasion of Lebanon in 1982, for example, articles in the mainstream Greek press referred to Israelis as "Nazis" and the "descendants of Hitler" (see, for a summary, *New York Times*, June 26, 1982). The reception of such statements in Greece, where little care has been taken to distinguish between Israel, Judaism, and Zionism, has been one that has helped to sustain and perpetuate various forms of anti-Semitism. The recent strengthening of diplomatic ties between Turkey and Israel has contributed to the lack of subtlety in distinguishing between the various issues—if the friend of one's friend is a friend, then one's enemy being friends with another enemy seems in this case to have made both doubly suspect—although the situation has improved somewhat in the past decade, since Prime Minister Konstantinos Mitsotakis recognized Israel in 1990.

Polls consistently show anti-Jewish attitudes among well over half the general population. This fact notwithstanding, the Greek government for the most part is exceedingly hesitant to acknowledge anti-Semitism as a real problem in Greek society. To some extent this is the intentional turning of a blind eye, but to a large degree it can be seen as the response of a society that is dramatically homogeneous in its ethnic makeup, and unskilled at dealing with the pressing issues of pluralism and tolerance. The very things that have historically given rise to Greek anti-Semitism—the linkage between Hellenism and Orthodoxy, and the ethnic homogeneity of Greek society—have, alas, also been the very things that have made Greece ill equipped to deal with it. It also has led to the widespread perception in Greece that its Jews somehow owe something to their Christian compatriots. In April 2002, for example, Theodoros Pangalos, former Greek foreign minister, wrote an op-ed piece in one of Athens' leading dailies, To Víma. Upset that Greece's Jews had not publicly protested the Israeli treatment of Palestinians during the siege of Jenin, he wrote that he was "sure that sitting in some of those [Israeli] tanks are the grandchildren of survivors saved from the Holocaust thanks to the solidarity of their Christian Greek fellow countrymen."[4] How are we to understand such a statement? As support for the Palestinians, yes. But more emphatically, as a suggestion that perhaps those Greek Jews shouldn't have been saved in the first place, as a reminder that they weren't ever really Greek to begin with, and as a claim that Greece's Jews today are, in the same way and for the same reason—by dint of being Jews—not really Greek either?

THE HOLOCAUST IN GREEK JEWISH MEMORY:
GREECE'S "RIGHTEOUS GENTILES"

That such attitudes are sadly prevalent among the Greek Christian populace renders particularly striking—and soberingly dignified—the fact that one of the primary modes through which the Greek Jewish community both affirms its Greekness and remembers the Holocaust is regular commemoration of the Greek Christians who came to the aid of Jews during World War II.

The Greek Jewish community has been persistent in its efforts to identify and thank the many Christians who helped Jews survive the war. The knowledge that many Christians risked their lives to help Jews "is the only thing that keeps me going," comments the granddaughter of a Salonikan stevedore who has spent her professional career documenting Greek Jewish history.[5] For those who were given assistance, the memory of the kindness of strangers is the only balm on the gaping wound of the Holocaust.

Koula Cohen Kofinas, who spent the occupation hiding in the mountains around Larissa, is adamant that the bravery of Christians who helped Jews not be forgotten. "There were some Christians who put their necks on the line or else nobody would be alive today. I tell it all the time and until the day I die I am going to tell it."[6] In 2000, an "Award Ceremony for Saviors" was held in Athens. Heinz D. S. Kounio, the president of the Salonikan Jewish Communal Assembly, wrote to the families of those honored. The Christians who helped Jews, he noted, "have lightened the dark and ugly paths of terror during th[e] hard times of the Nazi period."[7]

Some of the Christians who helped Jews did so for a price, but the vast majority were motivated simply by human decency. "They didn't think they were doing us a favor," comments a man who survived three years hiding in a woodshed outside of Athens. "They did it because they thought it was the right thing to do."[8] Many Jews were helped by complete strangers, whom they never saw again. In the thick of the occupation, Ilias Hadjis traveled from Larissa to Athens with his mother and brother. Just before the journey,

> a young fellow comes and tells my mother, "I know who you are. Don't worry, everything will be okay." My mother says to him, "I don't know who you are or what you're talking about." The guy leaves. When we get to Athens, we go outside the station and the young fellow comes back and says to my mother, "Do you know where you're going?" She says, "Yes, I'm going to the Athena Hotel and my husband will be there." "Do you have money to pay for the cab?" She says, "No, but my husband will." The guy calls a cab and asks the driver to take us there. The young man pays the fare and tells the driver, "You better take these people there," and disappears. Recently, I asked my mother if she knew the name of that person. She said, "I never saw him again in my life."[9]

Many of the Christians who helped Jews were members of the Left; after the war they were outcasts in Greek society. Greek Jewish organizations have been insistent that their bravery and kindness be remembered and acknowledged. Vasilis Rakopoulos, for example, a Greek doctor involved in the EAM, was imprisoned in the Mauthausen, Melk, and Ebensee concentration camps as a Communist. In Melk he served in the camp hospital, where he saved the lives of several Greek Jews, secretly giving them vital medication. From the Greek Jewish perspective, such individuals deserved accolades; the Greek government, however, was more interested in their political activities. "Rather than being honored as . . . war hero and humanitarian, he was persecuted as a Communist." Rakopoulos was imprisoned after the Greek Civil War and again by the Greek military junta that took over the country in 1967, spending "most of his adult life in concentration camps or prisons."[10] Many Greek Jewish survivors were grateful to the Left for its assistance during the occupation. Postwar Greek society's sharply hostile attitude toward former leftists was one of several factors that contributed to the increasing alienation from Greece felt by many Greek Jews in the 1940s and 1950s.

The commendation of Christians who came to the assistance of Jews has become an important feature of Greek Jewish memory of the Holocaust. The Kehila Kedosha Janina, now the only active Romaniote synagogue in the United States, has a permanent memorial to the "righteous Gentiles" who helped members of the Greek Jewish community; its museum has had exhibits on Christian saviors and the remarkable story of Zákinthos, and has collected oral histories from congregants who were hidden by Christians. The Jewish community in Greece has made great efforts to document stories of Christians' assistance. It is a tribute to the community that their Greek identity today is so bound up in the insistence that the fellow Greeks who helped them be remembered for having done so.

The Jewish history of Greece is really many histories of many places. Contemporary Greece encompasses territories that until recently were French, British, Italian, Venetian, and Ottoman. They've been influenced, in varying measure, by an array of different religious traditions and cultures. For centuries, some portions of present-day Greece were predominantly Jewish. Until 1943, Salonika, today's Hellenized Thessaloniki, was one of the largest Jewish cities in the world, and the only one where for generations Jews made up the majority population. Other regions, like the island of Crete in the south and the area around the town of Jannina in the northwestern mainland, were home to some of the oldest Jewish communities in the world, the Romaniotes, who settled in the region before the Christianization of the Balkan Peninsula.

The process by which these variegated peoples were nationalized—became Greek Jews—was never brought to completion. Over the course of the nineteenth and early twentieth centuries, Greek Jews underwent an uneven process of nationalization as Greece expanded at the expense of Ottoman territorial integrity. For the hundred years that Greece and the Ottoman Empire existed side by side, Greek Jews were anachronistically framed to a large extent by imperial categories. Just as the Greek nationalist platform of the Megáli Idéa framed all Greeks, both within and beyond the borders of the state, as part of the Greek polity, the religious groupings of the Ottoman Empire continued to draw Greek Jews into an Ottoman Jewish orbit long after the time of their actual incorporation into Greece. Full nationalization—that is, full identification on the part of Greek Jews themselves along with outside observers of them (although not, significantly, of non-Jewish Greeks) with the designator Greek Jews—was consequently belated, and came most rapidly only in diasporic (that is, non-Greek) contexts: New York City, Auschwitz-Birkenau, and Palestine/Israel. The paradox of the fact that Greek Jews fully came into existence only outside of Greece is one of the legacies of the awkward and still incomplete transition from a world of empires to one of nation-states.

The Jewish history of Greece is thus an elusive, phantom one—a history of a people who for the most part no longer exist, at least not in significant numbers in Greece. Greece today is home to less than five thousand Jews. The Greek Jew is both a modern and largely diasporic category. Several thousand more Greek Jews live outside of Greece—most in the United States and Israel, but also in parts of Africa and Latin America. As with any immigrant group, their identity as Greek is becoming diluted with each generation. Large numbers of Jews lived in Greece for only a brief time—from the conquest of Salonika in 1912 until the Nazi deportations of 1943–44, which killed close to 90 percent of Greek Jewry.

But the Jewish history of Greece is a phantom history in a second sense, too. The extent to which Jews in Greece are really Greek has long been debated, by Greek Jews, Greek Christians, and scholars alike. Since its inception in the early nineteenth century, Greece has largely defined itself as an Orthodox Christian country. Religious identity and national identity are closely intertwined; numerous Greek constitutions have asserted the inalienable link between Greekness and Orthodoxy. During what in Greek is called *Tourkokratía*, "Turkish rule," from the Ottoman conquests of the Balkans in the fourteenth and fifteenth centuries through the Greek War of Independence in 1812, the Ottomans grouped subject peoples according to religion, underscoring and propagating the centrality of Orthodoxy to Greek identity. When the Greek state was formed in the nineteenth century, advocates for a secular state based on post-Enlightenment concepts were in conflict with a view of Greekness that had the church as one of its strongest pillars. This side of the story is quite well-known to this day, attested to by the many Greek ultranationalist Web sites denouncing Greek Jews as Zionist traitors as well as the many anti-Semitic acts to which such ideologies give rise each year. The desecration of Greek Jewish cemeteries, defacing of Greek Jewish monuments, and vandalizing of synagogues remain sadly commonplace activities in Greece, as throughout Europe.

Judaism's self-characterization as a diasporic and locative religion, tied to a specific place from which all its members are declared somehow to have originally come, has also furthered the idea that Jews in Greece at the very least are simultaneously Greek and something else. The largest Greek Jewish group, the Sephardim of Salonika, who came to the Balkans after being ejected from Spain at the end of the fifteenth century, had a doubly exilic sense of themselves—both as Jews generally and more potently as expulsees of Sepharad. While Romaniote Jews spoke Greek, the Sephardim used Ladino, a medieval Jewish Spanish dialect passed down from the fifteenth century. The language set them apart from Greek speakers, but perhaps more critically it drew them toward their history in distant lands. For the most part Sephardim did not identify themselves as Greek in any way, and prior to the Greek conquest of Salonika in 1912, they overwhelmingly wished to remain part of the Ottoman Empire. Except for a few brief decades between the wars, Greek and Jew were regarded as separate categories by Greeks and Jews alike.

This dynamic has increased since the 1948 founding of Israel, which claims Jews around the world as part of its national constituency regardless of nationality. In Greece, where Israel enjoys less support than in perhaps any other Western country, Jews are often suspected of having divided national loyalties.[11] Greek anti-Semitism often conflates Jewishness and Zionism (as does Zionism itself). The conflict between Israel and Palestine has furthered tensions. Greece has long maintained strong and favorable ties to the Arab world; until the 1950s, a large and influential Greek diaspora lived in Alexandria, Egypt; many Palestinians are fellow Orthodox Christians (although the number has dwindled dramatically in past decades); and some of the Greek church's holiest sites are on Palestinian soil.

Paradoxically, though, what lies at the heart of much contemporary Greek anti-Semitic rhetoric are the strong similarities between the ways in which Greece and Israel define themselves as states that are at once democratic and somehow based on a national religion. A recent article in the Israeli newspaper *Ha-Aretz* reported that Israel had refused to allow citizens the option of registering as "Israelis" on the line marked "nationality" on the Israeli identity card. The State Prosecutor's Office argued that the category "does not reflect, is not suitable, and undermines the very principles under which the state of Israel was created [sic]." The state explained that "the dictionary definition of a nationality is 'a nation, a people; a large group of people of a joint origin, common destiny and history and usually a shared spoken language.'" The registration as Israeli would not reflect the "national and ethnic identity" of the petitioners. A group of Israelis that had petitioned for the right to register as Israeli countered, "The state is afraid that if it agrees to register an Israeli nationality, it will create a de facto separation between Jews abroad and Jews living in Israel as part of an Israeli nationality."[12]

Greece has recently had its own flap over identity cards, but with quite different causes and results. In response to a European Union directive, the Greek government agreed in 2000 to remove all references to an individual's religion from the state-issued identity card, contending that religion was a private matter of individual rights, not a legitimate national concern.[13] The Greek church, along with half a million protesters and three million signatories on a petition calling for a referendum, vociferously disagreed. When Prime Minister Kostas Simitis announced in May 2000 that new Greek identity cards would not specify the religion of the bearer, even on a voluntary basis, the church condemned the decision in terms reminiscent of early nineteenth-century attacks on the newly created division between church and state. "These changes are being put forward by neo-intellectuals who want to attack us like rabid dogs and tear at our flesh," protested Archbishop Christodoulos.[14] While the Greek government maintained that the inclusion of religion on the cards would lead to discrimination against members of religious minorities, its opponents countered that its removal was itself a form of discrimination against Orthodox Christians.[15] The protesters' position was that Orthodoxy was a critical component of Greekness—far from being a private matter, religious identity was one of the central markers of one's nationality.

In this parallel debate, the Israeli state and much of the Greek populace were in agreement that nationality and religious identity are coterminous. The agreement derives from a common source: just as Israel today wishes to establish a connection between Israel and Jews around the world, over the course of the nineteenth century Greece fought to maintain a connection between the Greek state and the Orthodox peoples who lived beyond its borders. In Greece, the territorial implications of this formulation have fallen away. Other than the addition of the Dodecanese Islands after World War II, Greece's borders have remained fixed since the 1920s. But the idea's core conceit has remained, as, for example, in the slogan made famous by Christodoulos, the chief prelate of the Greek Orthodox Church: "Greece means Orthodoxy." The Greece to which Christodoulos refers is more than a mere geopolitical construct. It is the Greek *ethnos*, a hazier and more emotionally connotative collectivity that takes its bearings from earlier formulations of collective identity.

This book has been about both: a place and a people. What, after all, is a Jewish history of Greece? There was no fixed entity called Greece until the close of the Greek War of Independence in 1833. When first established, its Jewish population likely numbered in the low thousands, if that. The Jewish history of Greece, then, is the history of the emergence and evolution of Greece as it affected the Jews who lived in the various territories that ultimately became part of it. It is the story of how, in the course of this evolution, some of the region's Jews were nationalized and became Greek while others rejected Greekness. Finally, it is the story of a postimperial, diasporic identity, one consolidated only during and after the 1940s, when Jews from around Greece came to regard themselves as a Greek Jewish collectivity. Greek territorial expansion, the rise of Zionism, the collapse of the Ottoman Empire after World War I, emigration, the Nazi occupation of Greece, the Holocaust, and the formation of Israel in 1948 are its many threads. Greek Jews were forged of several heterogeneous, diverse groups of people—and to diversity they have returned; along with Greece's small Jewish community, it is Salamonico, the congregants of the Kehila Kedosha Janina, and the singers of Radio Piraeus in Gush Dan who are the Greek Jews of today.

NOTES

CHAPTER 1. INTRODUCTION

1. Figures from Charles Moskos, *Greek Americans: Struggle and Success* (Engelwood Cliffs, NJ, 1980), 156.

2. For this and general population statistics in Epirus at the turn of the century, see M. V. Sakellariou, ed., Ηπείρου [Epirus] (Athens, 1997), 356–57.

3. Leon Colchamiro, Sabetz Menachem, Ezra Bacola, Avisay Gani, Elie Contente, Aaron Sadock, Samuel Benaderet, Menachem Josaphat, and Abraham Souice.

4. Jewish Community of Janina, Inc., Plaintiff, #4979–1914 C. County Clerk, New York County, filed and recorded October 30, 1914. Courtesy of the Kehila Kedosha Janina.

5. Arthus Foss, *Epirus* (Boston, 1978), 54–55.

6. Supreme Court, New York County. "In the Matter of the Application of the Society of Love and Brotherhood of Janina and the Israelite Community of Janina, Inc. and the United Brotherhood of Janina, Inc. for an order merging them into one corporation to be known as the United Brotherhood of Janina, Inc." February 1, 1926.

7. Καταστατικόν καί κανονισμοί της ηνωμένος Αδελφώτητος Ιωαννήνων, Νέας Ιόρκης, 1928 [Statutes and regulations of the incorporated Brotherhood of Jannina, New York]. The Harlem rabbi was Anzelo A. David; the Downtown Division's was Simon Asser. Courtesy of Kehila Kedosha Janina. In 1964, the Mapleton Synagogue in Bensonhurst, Brooklyn, was founded by two Romaniote Jews.

8. Bar Mitzfah Boys exhibit. Courtesy of the Kehila Kedosha Janina.

9. See, for context, Jonathan Boyarin, "Waiting for a Jew," in *Thinking in Jewish* (Chicago, 1996).

10. Among many others, see Hasia Diner, *Lower East Side Memories: A Jewish Place in America* (Princeton, NJ, 2000); Mario Maffi, *Gateway to the Promised Land: Ethnic Cultures on New York's Lower East Side* (New York, 1995); Ronald Sanders, *The Downtown Jews: Portraits of an Immigrant Generation* (New York, 1969).

11. "The History of Kehila Kedosha Janina." Courtesy of the Kehila Kedosha Janina.

12. *Jewish Heritage Report* 2, nos. 1–2 (Spring–Summer 1998); Elias V. Messinas, *The Synagogues of Salonika and Veroia* (Athens, 1997), 28–32.

13. *Kehila Kedosha Janina 75th Anniversary/Sisterhood of Janina 70th Anniversary* (May 19, 2002). Tribute book printed by Robert's Print Shop, Brooklyn, 2002. Courtesy of the Kehila Kedosha Janina.

14. Koula Cohen Kofinas, oral history recorded by the Kehila Kedosha Janina.

15. Personal communication, September 12, 2004.

16. Personal communication, Zvi Ben-Dor Benite, July 1999.

17. Personal communication, September 12, 2004.

18. See, for example, Bracha Rivlin, פנקס הקהילות: יון [Encyclopedia of Jewish Communities: Greece] (Jerusalem, 1998), published under the auspices of Yad Vashem as part of its national series documenting Jewish communities destroyed in the Holocaust.

19. Documentation of the Romaniotes is patchy. Of the various Jewish groups living in Greece, by far the greatest attention has been paid to the Sephardim. Histories of the Romaniotes include Rae Dalven, *The Jews of Jannina* (Philadelphia, 1990); Joshua Starr, *Romania: The Jewries of the Levant after the Fourth Crusade* (1949; repr., New York, 1980); Eftyhia Nahman, *Γιάννενα Ταξίδι στο Παρελθών* [Jannina Journey to the Past] (Athens, 1996).

20. Hayyim Yosef David Azulai and David Frankel, eds., ספר זרע אנשים: ובו זר"ע פסקים ועניּנים לאנשים רבים ["Seed of Man": Two Hundred and Seventy-Seven Rules and Issues for the Many] (Husiatyn, Ukraine, 1902); *Even ha'Ezer* 4 (National Library, Jerusalem, 8o2001, 318b). I use Minna Rozen's translation in "Individual and Community in the Jewish Society of the Ottoman Empire: Salonica in the Sixteenth Century," in *The Jews of the Ottoman Empire*, ed. Avigdor Levy (Princeton, NJ, 1994), 218. The Azulai and Frankel edition is in the Harvard College Hebrew collections.

21. See Aron Rodrigue, "The Ottoman Diaspora: The Rise and Fall of Ladino Literary Culture," in *Cultures of the Jews: A New History*, ed. David Biale (New York, 2002), 863–86.

22. See Avigdor Levy, *The Sephardim in the Ottoman Empire* (Princeton, NJ, 1992), 4.

23. Elias S. Artom and Humbertus M. D. Cassuto, eds., תקנות קנדיאה וזכרונותיה ספר: Statuta Iudaeorum Candiae. Eorumque Memorabilia [Registry of Protocols and Regulations of Kandia] (Jerusalem, 1943).

24. Ibid., 40.

25. I follow here David Biale's usage of the plural term: cultures. Biale asks, "Can we speak of one Jewish culture across the ages or only Jewish cultures in the plural, each unique to its time and place?" (*Cultures of the Jews*, xviii).

26. For various treatments of this topic, see, inter alia, Cemal Kafadar, *Between Two Worlds: The Construction of the Ottoman State* (Berkeley, CA, 1995); Maria Todorova, *Imagining the Balkans* (New York, 1997); Mark Mazower, *The Balkans: A Short History* (New York, 2000); Molly Greene, *A Shared World: Christians and Muslims in the Early Modern Mediterranean* (Princeton, NJ, 2000).

27. Stavroulakis, more than any one figure, established Greek Jewish history as a legitimate branch of Greek history, with a seemingly indefatigable range of abilities that have produced detailed paintings of Ottoman Jewish costume; collections of Greek Jewish recipes; books and articles on Greek Jewish history; and most recently, the more or less single-handed resurrection of the Etz-Ḥayyim synagogue in Hania, Crete. Stavroulakis is also the former director of the Jewish Museum of Greece, founded in 1977.

28. Athens is the largest, but active communities are also present in Salonika, Larissa, and Volos.

CHAPTER 2. AFTER INDEPENDENCE: "OLD GREECE"

1. Cited in *The Philhellene* 8, nos. 4–12 (April–December 1949): 4.

2. This violent outburst was unprecedented in the region, and its full context, scope, and origins have yet to receive scholarly treatment. Sources are also scarce.

3. Document 12:2, "The Holy Synod Anathematises the *Philiki Etairia*, March 1821," in *The Movement for Greek Independence, 1770–1821: A Collection of Documents*, ed. and trans. Richard Clogg (New York, 1976), 203–6.

4. Accounts of the event are slippery. In the second edition of his *Narrative of a Journey from Constantinople to England*, published in 1828, the Reverend Robert Walsh writes that various Jews by unfortunate happenstance were in the vicinity of the hanging and were pressed unwillingly into service by the vizier himself. See Yitzchak Kerem, "The Influence of Anti-Semitism on Jewish Immigration Patterns from Greece to the Ottoman Empire in the Nineteenth Century," in *Decision Making and Change in the Ottoman Empire*, ed. Caesar E. Farah (Kirksville, MO, 1993), 306, 306n3. In the fourth edition of Walsh's book (London, 1831), however, Walsh changes the account, writing that "the Jews volunteered their services to cast [the] body into the sea . . . and they dragged [the] corpse, by the cord by which he was hanged, through the streets with gratuitous insult" (10).

5. The famed Scottish historian George Finlay, resident in Greece just after the war, recounts the story and the way it was twisted so as to appear as a specifically anti-Christian event: "Th[e] odious task [of disposing of the patriarch's body] is rendered a source of horrid gratification to the Jewish rabble at Constantinople, by the intense hatred which prevails between Greeks and Jews throughout the east." George Finlay, *History of Greece* (Oxford, 1877), 6: 187–88.

6. Most noteworthy in this regard was the March and April 1822 campaign against Naoussa, which culminated in a massacre of its Greek Christian population (on April 8). "[M]any Jews, armed and thirsty for Christian blood, drag[ged] Christians out of the town, clubbed them on the head and when they fell slaughtered them as oxen." Two Salonikan Jewish merchants, Solomon Basaria and Aelion Isak, were commended for having offered "outstanding services to the Thessaloniki divan and the imperial army, helping in the extermination of the insurgents of Naoussa," in return for which they were allowed in July 1822 to pick from among the gathered plunder and have whatever they wished shipped to Thessaloniki. See John C. Vasdravelis, *The Greek Struggle for Independence: The Macedonians in the Revolution of 1821* (Thessaloniki, 1968), 123–24, 136.

7. David Brewer, *The Greek War of Independence* (Woodstock, NY, 2003), 112.

8. Cited in S. G. Howe, *An Historical Sketch of the Greek Revolution*, ed. George G. Arnakis (Austin, 1966), 88.

9. Percy Clinton Sydney Smythe Strangford, the sixth viscount Strangford (1780–1855), Public Record Office (PRO), Foreign Office (FO) 78/102, Report of the British Ambassador, November 21, 1821, Constantinople. Cited also in Kerem, "The Influence of Anti-Semitism," 307.

10. John Hartley, *Researches in Greece and the Levant* (London, 1831), 207. A Greek account is in Ἀπομνημονεύματα Ἠλίας Φωτάκου, ὑπασπιστῆ τοῦ

Κολοκοτρώνη [Memoirs of Ilias Fotakos, aide-de-camp of Kolokotronis] (n.p., 1858), 93.

11. For a summary of the topic, see Bracha Rivlin, פנקס הקהילות: יוון, [Encyclopedia of Jewish Communities: Greece] (Jerusalem, 1998), 124–25.

12. Hartley, *Researches in Greece and the Levant*, 207. Cited also in Kerem, "The Influence of Anti-Semitism," 307.

13. Abraham Galante, *Turcs et Juifs: Etude Historique, Politique* (Istanbul, 1932), 77.

14. For approximate population figures, see Rivlin, פנקס הקהילות, 101.

15. Ibid., 150.

16. See Nikos Stavroulakis, *The Jews of Greece* (Athens, 1990), 51; Bernard Pierron, *Juifs et Chrétiens de la Grèce Moderne* (Paris, 1976), 16.

17. Richard Clogg, *A Concise History of Greece*, 2nd ed. (Cambridge, UK, 2002), 261.

18. Markos Renieris, *Δοκίμιον φιλοσοφίας της ιστορίας* [Essay on the Philosophy of History] (Athens, 1841).

19. For useful demographic maps and charts, see J. M. Wagstaff, ed., *Greece: Ethnicity and Sovereignty, 1820–1994. Atlas and Documents* (London, 2002), 2–23. For commentary, see ibid., 57–59.

20. By an agreement signed in Constantinople, July 21, 1832, by representatives of Great Britain, France, Russia, and the Ottoman Empire.

21. For the full text of the conference, see Wagstaff, *Greece*, 151–59.

22. The loan was 60,000,000 francs, a sum first negotiated with the great powers by Leopold of Saxe-Coburg, son-in-law of King George IV of England, who had been first choice for monarch of Greece. After reconsidering his initial interest in the Greek post, Leopold became king of the Belgians instead. For detailed information on the debt, see Jon V. Kofas, *Financial Relations of Greece and the Great Powers, 1832–1862* (Boulder, CO, 1981); John A. Levandis, *The Greek Foreign Debt and the Great Powers, 1821–1898* (New York, 1944). Much of the information on the topic here is drawn from these sources.

23. *Banker's Magazine* (London) 63 (1863), 858. Cited in Levandis, *The Greek Foreign Debt and the Great Powers*, 15n70.

24. Edmund Lyons, cited in David H. Close, *The Origins of the Greek Civil War* (London, 1995), 2.

25. The Westernized Smyrniote Adamantios Korais (d. 1833), one of the leaders of the so-called Greek Enlightenment, singled out monks as the greatest evil of the church. John A. Petropulos, *Politics and Statecraft in the Kingdom of Greece, 1833–1843* (Princeton, NJ, 1968), 185.

26. For an overview of Greek education in the nineteenth century, and particularly of British travelers' views of it, see Helen Angelomatis-Tsougarakis, *The Eve of the Greek Revival: British Travellers' Perceptions of Early Nineteenth-Century Greece* (London, 1990), 118–45.

27. John Lloyd Stephens, *Incidents of Travel in Greece, Turkey, Russia, and Poland* (New York, 1840), 1:63.

28. Βασιλομένη δημοκρατία.

29. For summary accounts of Athenian Jewish history of the period, see Pierron, *Juifs et Chrétiens*; Stavroulakis, *The Jews of Greece*; Michael Molho, "La

Nouvelle Communauté Juive D'Athènes," in *The Joshua Starr Memorial Volume: Studies in History and Philology* (New York, 1953), 231–40.

30. Stavroulakis, *The Jews of Greece*, 100.

31. Molho, "La Nouvelle Communauté," 231.

32. Rivlin, פנקס הקהילות, 67–69. See also Molho, "La Nouvelle Communauté," 233; Stavroulakis, *The Jews of Greece*, 100.

33. Pierron, *Juifs et Chrétiens*, 23–24.

34. Rivlin, פנקס הקהילות, 67.

35. Molho, "La Nouvelle Communauté," 232; Stavroulakis, *The Jews of Greece*, 53.

36. PRO/FO 286/118, "Argent Déposé chez M. Pacifico par les communautés Israélites," August 14 (26), 1847, Athens.

37. Pacifico had been born on Gibraltar and could thus claim British citizenship.

38. PRO/FO 286/118, "Argent Déposé chez M. Pacifico," August 14 (26), 1847, Athens.

39. The stunningly beautiful jewelry of the ladies Pacifico included Portuguese gold filigree pieces, several brooches, gold bracelets weighing more than half a pound apiece, and a wide gold waistband adorned with two large rubies. PRO/FO 286/118, Athens, August 14 (26), 1847, "Bijoux, diamants, perles de Mme. D. Pacifico et de ses filles."

40. PRO/FO 286/118, "Conto della case del sig. Cavaliere Pacifico, ex console de Portogallo, suddito inglese," August 14 (26), 1847, Athens.

41. Pacifico (b. 1784) had served from 1835–37 as Portuguese consul in Morocco and thereafter as consul general in Greece. In most of his correspondence, Pacifico, who was Portuguese, uses the Gregorian calendar, adopted by most Catholic countries after it was established as the proper calendrical system by a papal bull in 1582. Protestant and Orthodox Christian countries took longer to change systems; in Greece, the Julian calendar was in use as late as 1924. Greek official documents on the Pacifico affair use the Julian calendar. In the 1800s, there was a twelve-day difference between the two (the Gregorian giving the earlier). Calculated algorithmically, Easter falls on a different date each year; within the Gregorian system, the cycle of Easter dates repeats only every 5,700,000 years. See http://www.smart.net/~mmontes/ec-cal.html

42. PRO/FO 286/113, Edmund Lyons to Colletti, April 26, 1847, Athens.

43. PRO/FO 286/118, David Pacifico to Edmund Lyons, October 12, 1847 (3 p.m.), Athens; PRO/FO 286/120 (no. 111), Palmerston to Edmund Lyons, December 18, 1847, London: PRO/FO 286/120, David Pacifico to Palmerston, October 8, 1847, Athens. The names of some of the participants are included in the legal briefs of the king's procurate. See PRO/FO 286/118 (no. 6927), "Le conseil della cour criminelle d'Athènes," August 15, 1847, Athens.

44. PRO/FO 286/118, David Pacifico to Edmund Lyons, September 9, 1847, Athens.

45. PRO/FO 286/117 (no. 75), Edmund Lyons to Palmerston, May 20, 1847, Athens.

46. PRO/FO 286/118, "Argent Déposé chez M. Pacifico," August 14 (26), 1847, Athens. The communities in question were in Florence, Paris, Trieste, and Venice.

47. PRO/FO 286/118, copies A-I, Athens (August 1, 1843, Trieste; April 23, 1844, Paris; May 16, 1843, Trieste; May 15, 1843, Trieste; May 11, 1843, Trieste; August 31, 1842, Venice; September 8, 1846, Florence; September 14, 1843, Florence).

48. See, for example, PRO/FO 286/118, David Pacifico to Edmund Lyons, third sheet, September 8 (20), 1847, Athens.

49. Thomas W. Gallant, *Modern Greece* (London, 2001), 108.

50. PRO/FO 286/117 (no. 75), Edmund Lyons to Palmerston, May 20, 1847, Athens.

51. PRO/FO 286/118 (no. 6927), "Le conseil della cour criminelle d'Athènes," 3, August 15, 1847, Athens.

52. PRO/FO 286/117, David Pacifico to Edmund Lyons, April 7, 1847, Athens.

53. "que D. Pasifico [sic] en payent les marguillers de cette église a reuissi d'empêcher qu'on brula l'image de Juda lequel faisait on par usage chaque année et brulais dans celtte paroisse au jour des Pâques." PRO/FO 286/118 (no. 6927), "Le conseil della cour criminelle d'Athènes," 3, August 15, 1847, Athens.

54. See, for example, Matthew 26:47ff.

55. Matthew 27:24-25. Translation is from *The New Oxford Annotated Bible with the Apocrypha* (Oxford, 1994). Pontius Pilate, the Roman governor in charge of the legal proceedings against Jesus, declares himself "innocent of this man's blood"; the people—Jews, the Gospel of John states—declare, "His blood be on us and on our children!" This unfortunate (and utterly decontextualized) passage was one scriptural basis and justification for acts of anti-Semitism toward Jews, particularly during Holy Week, when outbursts of anti-Jewish violence were tacitly supported by the authorities, who did little to intervene.

56. For an English version of the text, see George L. Papadeas, comp. and trans., *Holy Week-Easter: A New Translation* (South Daytona, FL, 1976).

57. Reaffirmed by Archbishop Damaskinos, when Greece was under German occupation.

58. See, for example, Pierron, *Juifs et Chrétiens*, 24-26 ("L'affaire Pacifico: acte anti-Semite ou affaire politique?").

59. Thomas W. Gallant, "Murder in a Mediterranean City: Homicide Trends in Athens, 1850-1936," *Journal of the Hellenic Diaspora* 24 (1998): 1-27.

60. Henry Martyn Baird, *Modern Greece: A Narrative of a Residence and Travels in That Country, with Observations on Its Antiquities, History, Language, Politics, and Religion* (New York, 1856), 74-75. Cited also in Gallant, *Modern Greece*, 111.

61. For a history of the Rothschild family and world banking in the period, see Niall Ferguson, *The House of Rothschild: Money's Prophets, 1798-1848*, vol. 1, and *The World's Banker, 1849-1999*, vol. 2 (New York, 1998-99). For Karl Marx's observations on the Rothschild loans to Greece, see *New York Daily Tribune*, March 15, 1853.

62. Britain paid Rothschild £23,570.17.1 against the Greek loan taken out in March. PRO/FO 286/117 (no. 76), Edmund Lyons to Palmerston, May 21, 1847, Athens; PRO/FO 286/113, Edmund Lyons to Coletti, May 16, 1847, Athens.

63. PRO/FO 286/118, December 31, 1844, Athens. The claim had been filed with the Portuguese government via the Athens notary on January 4, 1845. PRO/FO 286/124, David Pacifico to Edmund Lyons, February 22, 1848, Athens.

64. "loin de ma patrie dans un pays étranger." PRO/FO 286/118, March 30, 1847, Athens. Notary of Athens, Constantine Pittari, translated and copied by Papadiamandopoulos for the Chevalier D. Pacifico.

65. PRO/FO 286/120, Pacifico to Palmerston, October 8, 1847, Athens. "Les Ministres du Roi Othon ne sont pas encore assez avancés en civilization pour comprendre le Droit des Gens, les avantages de la tolerance religieuse."

66. See Jasper Godwin Ridley, Lord Palmerston (London, 1970).

67. "Outre que par ma position d'Israëlite, je suis inapt à agir comme un citoyen grec."

68. "C'est le Gouvernement Grec qui . . . m'a occasionné cet énorme prejudice." PRO/FO 286/124, Pacifico to Lyons, February 22, 1848, Athens.

69. See Sia Anagnostopoulou, Μικρά Ασία, 19ος αι.–1919 οι Ελληνορθόδοξες Κοινώτητες [Asia Minor, nineteenth century–1919: The Greek Orthodox Communities] (Athens, 1997).

70. Hertslet, no. 161, Arrangement between Great Britain, France, Russia, and Turkey, for the Definitive Settlement of the Continental Limits of Greece. Signed at Constantinople, July 21, 1832, and affirmed by the Conference of London, in its fifty-second protocol of August 30, 1832. Document 8 in Wagstaff, Greece, 160.

71. Cession of the Ionian Islands, 1864; acquisition of Thessaly and Arta, 1881; acquisition of Macedonia, Crete, Samos, Chios, Mytilini, and Lemnos, 1913; cession of eastern Thrace and portions of Asia Minor by the unratified Treaty of Sevres, 1920 (lost in 1922); cession of central/western Thrace by Treaty of Sevres, 1920, ratified by Treaty of Lausanne, 1923; cession of Dodecanese by Italy, 1947.

72. For the entire speech, see Yiannis Kordatos, Ιστορία της Νεότερης Ελλάδας [History of Modern Greece] (Athens, 1957–58), 303–6.

73. Tasos Vournas, ed., Μακρυγιάννη—Απομνημονεύματα [Makriyannis—Memoirs] (Athens, 1972), 219.

CHAPTER 3. "NEW GREECE": GREEK TERRITORIAL EXPANSION

1. For a complete account of the siege, 1912–13, see Guy Chantepleure, La Ville Assiegée: Janina—Octobre 1912–Mars 1913 (Paris, n.d.).

2. Arthur Foss, Epirus (Boston, 1978), 34.

3. Interview with J. Tsito, in Annette B. Fromm, "We Are Few: Folklore and Ethnic Identity of the Jewish Community of Ioannina, Greece" (PhD diss., Indiana University, 1992), 32.

4. A.A.D. Seymour, "Caveat Lector: Some Notes on the Population Figures of the Jewish Communities of the Ionian Islands: II," Bulletin of Judaeo-Greek Stud-

ies 15 (Winter 1994): 33. For the year 1802, Seymour uses figures from the Ionian government census, in J. Hennen, *Sketches of the Medical Topography of the Mediterranean* (1830); for 1891, he references a citation of Rabbi Solomon Levi of Zante. See Seymour, "Caveat Lector," 33–39. See also Bracha Rivlin, הקהילות: יוון פנקס [Encyclopedia of Jewish Communities: Greece] (Jerusalem, 1998), 353.

5. Rivlin, פנקס הקהילות, 353.

6. Ibid., 131

7. For this and general population statistics in Epirus at the turn of the century, see M. V. Sakellariou, ed., *Ηπείρου* (Athens, 1997), 356–57.

8. Foss, *Epirus*, 54–55.

9. Rivlin, פנקס הקהילות, 194.

10. See Annette B. Fromm, "A Ritual Blood Libel in Northwestern Greece," in *From Iberia to Diaspora: Studies in Sephardic History and Culture*, ed. Yedida K. Stillman and Norman A. Stillman (Leiden, 1999), 48–57.

11. For a summary account, see Pearl L. Preschel, "The Jews of Corfu" (PhD diss., New York University, 1984), 87ff.

12. *Jewish Chronicle*, May 22, 1891.

13. Rivlin, פנקס הקהילות, 353. See also Cecil Roth, *Venice* (Philadelphia, 1930), 330.

14. Rivlin, פנקס הקהילות, 353.

15. Roth, *Venice*, 329.

16. *Jewish Chronicle*, November 19, 1858, 2.

17. Shmuel Sardas, who attended his local demotic school on Corfu in the 1920s, recalls it as one of the happiest periods of his life, but also remembers that "on the eve of Easter, the Christians used to throw clay pots out of their windows, shouting, 'On the heads of the Jews!'" Shmuel Refael, ed., בנתיבי שאול: יהודי יוון בשואה [Road to Hell: Greek Jews in the Holocaust] (Jerusalem, 1988), 392–93.

18. George Orkney, *Four Years in the Ionian Isles* (London, 1864), 2:47–48.

19. ספונות [*Sefunot*] 1, "מקורות לקורות קהילת קורפו" [Sources for the History of the Jewish Community of Corfu], in ספונות, book 1 (Jerusalem, 1957), 311–13.

20. Ibid., 313.

21. Anxieties about the proximity of Jews and Orthodox Christians is also reflected in the story of the Jewish maiden Rachel Vivante, who in April 1776 was snatched from her home by the Greek Orthodox Spyridonas Voulgaris. See A. Ch. Tsitsas, *Η Απαγωγή της Ραχήλ Vivante* [The abduction of Rachel Vivante] (Corfu, 1993).

22. Cited in Meir Benayahu, בניהו, רבי אליהו קפשאלי, איש קנדיאה: רב, מנהיג והיסטוריון מאיר [Rabbi Eliyahu Kapsali, Man of Candia: Rabbi, Leader, and Historian] (Tel Aviv, 1983), 110. Benayahu points out that there are several other *takkanot* to similar effect.

23. Roth, *Venice*, 316.

24. Cited in Salo W. Baron, "Jewish Immigration and Communal Conflicts in Seventeenth-Century Corfù," in *The Joshua Starr Memorial Volume: Studies in History and Philology* (New York, 1953), 171–72.

25. "מקורות לקורות קהילת קורפו," 306–9.

26. See, for example, "Universita Italiana della Sinagoga Corfiota Aborigine contro Universita della Sinagoga Corfiota Greca" (Columbia University Special

Collections, X893-28-v.3-no.6), reproduced in Baron, "Jewish Immigration and Communal Conflicts," 173–78.

27. "ספונות [Sefunot] 15, "אגרות אל שלוחי ארץ ישראל בקורפו" [Epistles concerning Messages from the Jews of Corfu], in ספונות, book 15 (Jerusalem, 1981), 56–58.

מלאו מתנינו חלחלה צירים אחזונו כצירי יולדה, את נשף חשקנו שם לנו לחרדה
כיקול המולה שמענו קול מלחמה במחניכם קדש . . . לחלל שם שמים וליקר שם
האלילים. להשכיחם ללכת אחרי אלקים אחרים לעבדם בטהרה ולהעלות לאפם הזמורה
ולמסור איש את ממון חבירו ביד דינים זרים שנצטוו במר"ה . . . בעת שהתורה
חוגרת שק עליהם על כי בילדי נכרים ישפיקו בחציריהם ובטירותם

28. See Ḥayyim Mizraḥi and David Benveniste, "רבי יהודה ביבס וקהילת קורפו בזמנו" [Rabbi Yehudah Bivas and the Community of Corfu during his Lifetime], in ספונות, book 2 (Jerusalem, 1957), 303–30.

29. Ibid., 318–21. In return for his services, the rabbi would be given a salary and moving fees, along with a free home "adjacent to the *midrash* [school], and from the windows of which you can see the sea."

30. Ibid., 305, 315–18.

31. Baron, "Jewish Immigration and Communal Conflicts," 171.

32. Richard Ridley Farrer, *A Tour in Greece, 1880* (Edinburgh, 1882), 5, 7, 9.

33. For detailed information on Sicilian Jewish migration to the eastern Mediterranean, see Attilio Milano, *Storia degli ebrei italiani nel levante* (Florence, 1949).

34. In the papers of Doge Leonardo Donato (1535–1612) we find the following breakdown of the Corfiote Jews, who he numbered in 1607 at five hundred: "Gli ebrei [sono] divisi in tre classi, e cioè: Antiche di Corfù; una seconda di non molto antichi, ed una terza, composta di circa novanta individui Portoghesi, Spagnoli, Levantini, che attendi ai traffici." In L. A. Schiavi, "Gli ebrei in Venezia e nelle sue colonie," *Nuova Antologia* 47 (1893): 486. Cited in Seymour, "Caveat Lector," 34.

35. I use Minna Rozen's translation in "Individual and Community in the Jewish Society of the Ottoman Empire: Salonika in the Sixteenth Century," in *The Jews of the Ottoman Empire*, ed. Avigdor Levy (Princeton, NJ, 1994), 218.

36. Rashdam, *Responsa, Oraḥ Ḥayyim* 36. Cited in Rozen, "Individual and Community," 224.

37. For an overview of the topic, particularly as reflected in rabbinic writings, see Bracha Rivlin, "The Greek Peninsula, a Haven for Iberian Refugees: Effects on Family-life" [sic], in *The Jewish Communities of Southeastern Europe: From the Fifteenth Century to the End of World War II*, ed. I. K. Hassiotis (Thessaloniki, 1997), 443–52.

38. S. Rosanes, דברי ימי ישראל בתוגרמה [The History of the People of Israel in Turkey] (Tel Aviv, 1930), 1:216.

39. At the time of the expulsions, Kavála's dominant Jewish group was not Romaniote but Hungarian Ashkenazic. Ashkenazim and Romaniotes alike rapidly became assimilated into Sephardic *minhag*.

40. Giomtov Giakoel and Fransiski Ampatzopoulou, Απομνημονεύματα *1941–1943* [Memoirs, 1941–1943] (Thessaloniki, 1993), 12–13.

41. Rebecca Fromer, *The House by the Sea: A Portrait of the Holocaust in Greece* (San Francisco, 1998), 55.

42. Chantepleure, *La Ville Assiegée*, 105.

43. Joshua Eli Plaut, *Greek Jewry in the Twentieth Century, 1913–1983* (Madison, NJ, 1996), 28.

44. Rivlin, פנקס הקהילות, 67.

45. Cited in ibid., 71–72.

46. Nikos Stavroulakis, *The Jews of Greece* (Athens, 1990), 54.

CHAPTER 4. SALONIKA TO 1912

1. Mark Mazower, *Salonica, City of Ghosts: Christians, Muslims, and Jews, 1430–1950* (London, 2004).

2. Flory Jagoda, "Memories of Sarajevo: Judeo-Spanish Songs from Yugoslavia" (Global Village Music, 1991).

3. On the use of responsa literature as a historical source, see Marc D. Angel, "The Responsa Literature in the Ottoman Empire as a Source for the Study of Ottoman Jewry," in *The Jews of the Ottoman Empire*, ed. Avigdor Levy (Princeton, NJ, 1994), 669–85.

4. See Minna Rozen, "Individual and Community in the Jewish Society of the Ottoman Empire: Salonika in the Sixteenth Century," in *The Jews of the Ottoman Empire*, ed. Avigdor Levy (Princeton, NJ, 1994), 215–73, esp. 216–24, "The Trials of Immigration and Reorganization."

5. A good overview is provided by Rivka Koen. "Some Problems of Religious and Social Absorption of Ex-Marranos by Greek Jewry under Ottoman Rule after 1536," in מליסבון לשלוניקי וקושטא [From Lisbon to Salonika and Constantinople], ed. Zvi Ankori (Tel Aviv, 1988), 11–26.

6. Rashdam was the head of the congregation Gerush Sefarad (lit., "the expulsion from Sepharad") in Salonica. Rashdam, *Responsa, Even ha-'Ezer* 112.

(שכל האנוסים הבאים לעשות תשובה, כשם שמחזיקים
אותו שאביו מישראל, כך מחזיקים אותו שאמו מישראל ואינה גויה)

7. Ibid., 127.

(שהרי עקרו דירתם ממלכות פורטוגל לבוא תוגרמה
גם כי קצת מהם משתקעים באנקורה ובפלנדריה, הם מיעוטא דמיעוטא ואיכא למימר
דהוו ליה כיהודים כשרים, כיון דהרהרו תשובה בלבם)

8. Rival, *Responsa* 8–1, 22.

9. Rashdam, *Responsa, Yoreh de'ah* 199.

10. Moses Ben-Baruch Almosnino, *Cronica de los Reyes Otomanos*, ed. and trans. Pilar Romeu Ferré (Barcelona, 1998).

11. For an overview, see Theodore George Tatsios, *The Megali Idea and the Greek-Turkish War of 1897: The Impact of the Cretan Problem on Greek Irredentism, 1866–1897* (Boulder, CO, 1984).

12. P. K. Enepekidis, *Η Θεσσαλονίκη στα Χρόνια 1875–1912* [Thessaloniki in the Years 1875–1912] (Thessaloniki, 1988), 13.

13. Antonis Liakos, *L'Unificazione Italiana e la Grande Idea: Ideologia e azione dei movimenti nazionali in Italia e in Grecia, 1859–1871* (Florence, 1995), 87.

14. John S. Koliopoulos, *Brigands with a Cause: Brigandage and Irredentism in Modern Greece, 1821–1912* (Oxford, 1987), 323.

15. G. Aspreas, *Πολιτική Ιστορία της Νεότερας Ελλάδος* [Political History of Modern Greece] (Athens, 1960–69), 2:250–72.

16. Koliopoulos, *Brigands with a Cause*, 325.

17. Rena Molho, "The Zionist Movement in Thessaloniki, 1899–1919," in *The Jewish Communities of Southeastern Europe: From the Fifteenth Century to the End of World War II*, ed. I. K. Hassiotis (Thessaloniki, 1997), 330.

18. PRO/FO 78/4828, Alliance Israélite of Salonika to Israelite Community of Corfu, May 13 (25), 1897, Salonika.

19. *Ακρόπολη* [Acropolis], May 2, 1897.

20. PRO/FO 78/4828, Israelite Community of Corfu, May 4 (16), 1897.

21. PRO/FO 78/4828, Alliance Israélite of Salonika to Israelite Community of Corfu, May 13 (25), 1897, Salonika.

22. PRO/FO 78/4828, J. E. Blunt to the British Secretary of State for Foreign Affairs, June 9, 1897, Salonika.

23. Salonika was an exception in the region. Zionism was also more popular in Istanbul. See Esther Benbassa, "Presse d'Istanbul et de Salonique au service du sionism (1908–1914): les motifs d'une allegeance," *Revue Historique* 560 (October–December 1986): 337–65. On Salonikan Zionism, see Acher Moissis, "El mouvimiento sionista en Salonique i en las otras sivdades de Grecia," in דבלקן זכרון שלוניקי: גדולתה וחורבנה של ירושלים [Memoir of Salonika: The Glory and Destruction of Jerusalem of the Balkans], ed. David Recanati (Tel Aviv, 1972), 1:44–48; Molho's excellent overview, "The Zionist Movement in Thessaloniki," 327–50.

24. The best and most comprehensive recent account of Jewish Salonika in the late nineteenth and early twentieth centuries is Rena Molho, *Οι Εβραίοι της Θεσσαλονίκης 1856–1919* [The Jews of Thessaloniki, 1856–1919] (Athens, 2001). A number of other works have covered the topic as well. Salonika in the late Ottoman period is one of the best-researched topics in the rapidly growing literature on south Balkan and Greek Jewry.

25. David Florentin, cited in E. Kostas Skordyles, "Reactions juives a l'annexion de Salonique par la Grèce (1912–1913)," in *The Jewish Communities of Southeastern Europe: From the Fifteenth Century to the End of World War II*, ed. I. K. Hassiotis (Thessaloniki, 1997), 502. "Salonique serait comme un coeur qui cesserait de batter. Il serait comme une tête qui serait tranchée de son corps dépecé."

26. See Douglas Dakin, *The Greek Struggle in Macedonia, 1897–1913* (Thessaloniki, 1966).

27. For a superb overview of the period, see Yitzchak Kerem, "The Europeanization of the Sephardic Community of Salonika," in *From Iberia to Diaspora: Studies in Sephardic History and Culture*, ed. Yedida K. Stillman and Norman A. Stillman (Leiden, 1999), 59–74.

28. Esther Benbassa and Aron Rodrigue, *The Jews of the Balkans: The Judeo Spanish Community, 15th to 20th Centuries* (Oxford, 1995), 70.

29. Natasha Gaber and Aneta Joveska, "Macedonian Census Results—controversy or Reality?" *South-East Europe Review* 1 (2004): 99–100.

30. Cited in "מי ומי המפחדים?" [Who Are Those Who Are Afraid?], המבשר [*Ha-Mevasser*] 2, no. 1 (19 Tevet 1911): 3.

31. For an overview of the Greek Jewish press in the period, see Solomon Ruben-Mordechai, "Ο Εβραϊκός Τύπος στη Θεσσαλονίκη και Γενικότερα στην Ελλάδα" [The Jewish Press in Thessaloniki and in Greece Generally], Χρονικά [*Hronika*] 1 (June 1978): 1–20.

32. Thessaloniki Court of the First Instance, File 7; Archives de l'Alliance Israélite Universelle (AAIU), GR.I/G.3, May 19, 1919. Cited in Molho, "The Zionist Movement in Thessaloniki," 329n11.

33. David Matalon, secretary of the Salonika's "New Club," Sabbath sermon. In "מכתבים מסלוניקי" [Letters from Salonika], המבשר 2, no. 2 (January 26, 1911): 19.

34. AAIU, GR.I/G.3, May 19, 1919. Cited in Molho, "The Zionist Movement in Thessaloniki," 330.

35. "מי ומי המפחדים?" [Who Are Those Who Are Afraid?] המבשר 2, no. 1 (January 19, 1911): 3. The piece produced an episode of intrigue; the issue of *Hak* in which the admonition appeared was sent anonymously to various suspected pro-Zionist leaders, with the article highlighted in blue pencil. General consensus was that this act was the work of anti-Zionist Jews who wished to besmirch the good name of others by tarring them with the brush of Zionism.

36. Chief Rabbi Yaacov Meir, member of the Kehila Committee; Shmuel Di Modiano; Yosef Naar; Yitzhak Florentin; Benico Saltiel. המבשר 2, no. 10 (March 24, 1911).

37. Kostas Tomanas, Οι Κάτοιχοι της Παλιάς Θεσσαλονίκης [The Residents of Old Thessaloniki] (Thessaloniki, 1992), 43.

38. Bernard Pierron, *Juifs et Chrétiens de la Grèce Moderne* (Paris, 1976), 71.

39. See, for a summary, Justin McCarthy, *The Ottoman Peoples and the End of Empire* (New York, 2001), 53–60; see also Mazower, *Salonica*, 284–85.

40. See Moissis, "El mouvimiento sionista en Salonique," 1:44–48. Cited also in Molho, "The Zionist Movement in Thessaloniki," 331n18.

41. Jacob M. Landau, *Tekinalp, Turkish Patriot, 1883–1961* (Leiden, 1984), 10.

42. Numerous studies of the period have been published. See, inter alia, Avigdor Levy, trans. and ed., *The Jews of the Ottoman Empire* (Princeton, NJ, 1994); Avram Galante, *Histoire des Juifs de Turquie*, 8 vols. (repr., Istanbul, 1985); I.-S. Emmanuel, *Histoire des Israélites de Salonique*, vol. 2 (Paris, 1936); Benbassa and Rodrigue, *The Jews of the Balkans*. A superb synthetic treatment of Jewish-Muslim-Christian interactions in Salonika, 1600–1800, is found in Mazower, *Salonica*, 66–139.

43. AAIU, GR.I/G.3, July 4, 1919. Cited in Molho, "The Zionist Movement in Thessaloniki," 349.

44. Pierron, *Juifs et Chrétiens*, 72.

45. Molho, "The Zionist Movement in Thessaloniki," 329n12.

46. E. Schaap, "Back to Zion or Mesopotamia," *African Times and Orient Review*, September 1912, 87–89.

47. AAIU, GR.I/C.1–52, January 27, 1910. Cited also in Molho, "The Zionist Movement in Thessaloniki," 331n22.

48. For a summary overview of Maccabi, which evolved over time from being a Zionist organization to a nondenominational youth club, see Molho, "The Zionist Movement in Thessaloniki," 333–38.

49. For an overview, see Lily Macrakis, "Eleftherios Venizelos in Crete, 1864–1910: The Main Problems," in *New Trends in Modern Greek Historiography*, ed. Lily Macrakis and P. N. Diamandouros (New Haven, CT, 1982).

50. The Goudi coup was a military revolt that erupted in August 1909, largely in reaction to the Greek government's evident inability to create either domestic stability or international success. It was also a reaction to the Young Turk rebellion of 1908, which abolished the sultanate, restored constitutional rights to Ottoman subjects—an event that brought about change in Salonika's Jewish community—and provided Bulgaria the opportunity to declare full independence and Austria to annex Bosnia-Herzegovina. On Crete, it led to an ultimately successful uprising and swept Venizélos into power with the assistance of a military league (στρατιωτικός σύνδεσμος) that overrode the Greek government's dithering in the face of these events.

51. For a full account of 1901, see Phillip Carabott, "Politics, Orthodoxy, and the Language Question in Greece: The Gospel Riots of November 1901," *Journal of Mediterranean Studies* 3, no. 1 (1993): 117–38. The 1902 riots, the "*sandika*" demonstrations, followed the fall elections that year. See Thomas W. Gallant, *Modern Greece* (London, 2001), 120.

52. Victor Papacosma, *The Military in Greek Politics: The 1909 Coup d'Etat* (Kent, OH, 1977), 16.

53. On the anti-Hellenizing efforts, see *Le Fait de la Semaine* 6e, no. 12 (February 9, 1918); *Les Persecutions anti-helleniques en Turquie D'après les rapports officials des Agents diplomatiques et consulaires* (Paris, 1918), 2.

54. Mark Mazower, review of *A Sephardi Life in Southeastern Europe: The Autobiography and Journal of Gabriel Arié, 1863–1939*, ed. Esther Benbassa and Aron Rodrigue, *Bulletin of Judaeo-Greek Studies* 24 (Summer 1999): 19. This was not unlike the situation in which many Sephardic Jews found themselves on their arrival in Israel in the early 1950s.

CHAPTER 5. BECOMING GREEK: SALONIKA 1912–23

1. PRO/FO 608/44, W.L.C. Knight (Acting British Consul General in Salonika) to British Secretary of State, April 29, 1919.

2. Ibid.

3. G. Ward Price, *The Story of the Salonika Army* (New York, 1918), 93.

4. PRO/FO 371/1778, David Alexander and Claude G. Montefiore to Saint James Palace, December 17, 1912, London.

5. PRO/FO 608/44, W.L.C. Knight to British Secretary of State.

6. PRO/FO [53264], no. 313, enclosure 2, Consul-General Harry H. Lamb to Sir G. Lowther, December 3, 1912, Salonika.

7. E. Kostas Skordyles, "Reactions juives à l'annexion de Salonique par la Grèce (1912–1913)," in *The Jewish Communities of Southeastern Europe: From the Fifteenth Century to the End of World War II*, ed. I. K. Hassiotis (Thessaloniki, 1997), 510.

8. David Florentin to Zionist Organization Committee, December 15, 1912, Berlin. Cited in David Recanati, ed., זכרון שלוניקי: גדולתה וחורבנה של ירושלים דבלקן [Memoir of Salonika: The Glory and Destruction of Jerusalem of the Balkans] (Tel Aviv, 1972), 1:324–26.

9. AAIU, GR.XVII/E.202, May 6, 1914.

10. For an overview of the effort to internationalize Thessaloniki, see N. M. Gerber, "An Attempt to Internationalize Salonika: 1912–1913," *Jewish Social Studies* 17, no. 4 (October 1955).

11. Mark Mazower, *Salonica, City of Ghosts: Christians, Muslims, and Jews, 1430–1950* (London, 2004), 300.

12. AAIU, GR.I/C.50, May 29, 1912.

13. *Le Radical* (Paris), December 14, 1912, cited in Rena Molho, "The Zionist Movement in Thessaloniki, 1899–1919," in *The Jewish Communities of Southeastern Europe: From the Fifteenth Century to the End of World War II*, ed. I. K. Hassiotis (Thessaloniki, 1997), 340n59.

14. Ibid., 340.

15. Socrates A. Xanthaky and Nicholas G. Sakellarios, trans., *Greece in Her True Light, Her Position in the World-Wide War as Expounded by El. K. Venizelos, Her Greatest Statesman, in a Series of Official Documents* (New York, 1916), 270.

16. Μακεδονία [*Makedonía*] and Εμπρός [*Embrós*], October 28, 1912.

17. PRO/FO 608/44, W.L.C. Knight to British Secretary of State.

18. P. K. Enepekidis, *Η Θεσσαλονίκη στα Χρόνια 1875–1912* (Thessaloniki, 1988), 351.

19. Ibid., 352.

20. AAIU, GR.I/D.3, November 23, 1912 and December 12, 1912; GR.I/C.34–51, November 24, 1912. Cited also in Molho, "The Zionist Movement," 338n46.

21. Much as the French are today, in response to Israeli suggestions that France is not a safe place for Jews.

22. PRO/FO 608/44, W.L.C. Knight to British Secretary of State.

23. *First Memorandum to King Constantine*, January 11 (24), 1915, Athens, in Xanthaky and Sakellarios, *Greece in Her True Light*, 25–40. Venizélos's subsequent September 1915 speeches to the House of Representatives, included in the same volume, give a vivid picture of the vicious debates in the period between the Venizelists and anti-Venizelists. Gounaris, along with the representatives from Patras and Corfu, were particularly vituperative in their condemnations of Venizélos's proposed policy. The speeches also make clear the proportions of the constitutional crisis that arose from the conflict.

24. Venizélos and Coundouriotes, Proclamation to the Greek People, September 14, 1916, Hania, Crete, in Xanthaky and Sakellarios, *Greece in Her True Light*, 201, 203.

25. Viscount Northcliffe, Introduction to G. Ward Price, *The Story of the Salonika Army*, by G. Ward Price (New York, 1918), x.

26. Venizélos to the Greek House of Representatives, September 21, 1915, in Xanthaky and Sakellarios, *Greece in Her True Light*, 67, 91.

27. Venizélos to the Greek House of Representatives, October 15, 1915, in Xanthaky and Sakellarios, *Greece in Her True Light*, 191–92.

28. Venizélos to the Greek House of Representatives, September 28, 1915, in Xanthaky and Sakellarios, *Greece in Her True Light*, 150.

29. Much as the king and Venizélos were to blame for the increasingly volatile domestic situation, the great powers had even greater responsibility. Their endless interventions in the region, coupled with a poor understanding of its internal dynamics, severely undermined Greece's ability to develop viable political structures. The West didn't grasp fully the implications of expansionist ideologies such as the Great Idea, which they dismissed as silly without taking into account their tremendous power to motivate small and impoverished Balkan nations to take up arms against one another. For a detailed and heavily researched account of ties between Greece and the West in the period, see George F. Leon, *Greece and the Great Powers, 1914–1917* (Thessaloniki, 1974).

30. Harold Lake, *In Salonica with Our Army* (London, 1918), 269–70.

31. Kafantares, spokesman for the majority on the answer to the kings' speech, August 11 (24), 1917; Repoules, minister of the interior, August 12 (25), 1917; Politis, minister for foreign affairs, August 13 (26), 1917; Eleuthérios Venizélos, president of the council, August 13 (26), 1917; Stratos, member for Aetolia-Acarnania, August 10 (23), 1917; Rallis, member for Attico-Boeotia, August 12 (25), 1917. In *The Vindication of Greek National Policy, 1912–1917: A Report of Speeches Delivered in the Greek Chamber, August 23 to 26, 1917* (London, 1918).

32. On the arrival of Allied troops, see Recanati, זכרון שלוניקי, 1:205–6.

33. Price, *The Story of the Salonica Army*, 7–8.

34. Cited in ibid., 223.

35. In Recanati, זכרון שלוניקי, 1:205.

36. Edward Lear, *Journals of a Landscape Painter in Greece and Albania* (London, 1988), 20.

37. Cited in Recanati, זכרון שלוניקי, 1:205.

38. See *Palestine and Transjordan* 1, no. 22 (October 31, 1936): 1–2; *New Judaea* 21, nos. 1–2 (October–November 1944): 24. In the mid-1930s, Jews who used the Jaffa port rather than Tel Aviv were beaten by other Jews for supporting the Arab economy and undermining the Jewish one. See *Palestine and Transjordan* 1, no. 42 (March 27, 1937): 4.

39. PRO/FO 608/44, W.L.C. Knight to British Secretary of State.

40. Price, *The Story of the Salonica Army*, 81–84.

41. Cited in Recanati, זכרון שלוניקי, 1:206.

42. Ibid.

43. For a detailed statistical account of the fire's effects and its aftermath, see Rena Molho, *Οι Εβραίοι της Θεσσαλονίκης 1856–1919* (Athens, 2001), 120ff. See also Bernard Pierron, *Juifs et Chrétiens de la Grèce Moderne* (Paris, 1976), 91–115.

44. August 18–19 by Greek calendar.

45. Numbers vary by about ten thousand on either end of the range. Pierron's are used here as a guideline.

46. Price, *The Story of the Salonica Army*, 87.

47. Cited in Recanati, זכרון שלוניקי, 1:209.

48. Cited in Price, *The Story of the Salonica Army*, 88.

49. *L'Echo de France*, August 22, 1917; Joseph Nehama, *Histoire des Israélites de Salonique* (1935; repr., Thessaloniki, 1978), 7:764.

50. Cited in Recanati, זכרון שלוניקי, 1:207.

51. Pierron, *Juifs et Chrétiens*, 92.

52. Recanati, זכרון שלוניקי, 1:138.

53. Law 823, September 2, 1917.

54. Pierron, *Juifs et Chrétiens*, 92. See also contemporary issues of *La Epoca* and *El Pueblo*.

55. Law 1122, January 4, 1918; Law 1394, May 3, 1918.

56. Recanati, זכרון שלוניקי, 1:209.

57. David Benveniste, יהודי שאלוניקי בדורות האחרונים: הליכות חיים, מסורת וחברה [The Jews of Salonika in Latter Generations: Customs, Tradition, and Society] (Jerusalem, 1973), 42–43.

58. Nehama, *Histoire des Israélites*, 7:769.

59. Enrique Saporta y Beja, *En Torno de la Torre Blanca* (Paris, 1982), cited in Pierron, *Juifs et Chrétiens*, 95.

60. *L'Echo de France*, August 22, 1917.

61. Pierron, *Juifs et Chrétiens*, 95.

62. *L'Echo de France*, August 22, 1917. "On peut dire que tout Salonique est en flammes. Le spectacle est terrifiant."

63. Recanati, זכרון שלוניקי, 1:207–8.

64. For various figures, see Nehama, *Histoire des Israélites*, 7:765–67; Molho, *Οι Εβραίοι της Θεσσαλονίωϋ 1856–1919*, 121 (on the value of lost property); Pierron, *Juifs et Chrétiens*, 97. Again, figures vary considerably. Pierron has analyzed the various figures available and seems of the numerous accounts to be the most judicious in those he cites.

65. A list and valuation of Jewish community property lost in the fire is in the AAIU, GR. II.C. 54, "response au questionnaire de la mission Hoover."

66. Molho gives the figure as 450 torah scrolls in *Οι Εβραίοι της Θεσσα λονίωϋ 1856–1919*, 121.

67. Recanati, זכרון שלוניקי, 1:207.

68. Molho, *Οι Εβραίοι της Θεσσαλονίκης 1856–1919*, 121.

69. Nehama, *Histoire des Israélites*, 7:767–768.

70. Recanati, זכרון שלוניקי, 1:208–9.

71. *Il Pueblo*, December 10, 1917, and March 5, 1918.

72. PRO/FO 608/44, W.L.C. Knight to British Secretary of State.

73. *L'Opinion* (Salonika), November 1–17, 1917. This French-language Jewish paper, the mouthpiece of Venizelist and other "assimilationist" Jews, was published by Mentech Ben Sandci.

74. At the time of the fire, there were at least nine Ladino-language newspapers in circulation; four were created in response to it, among these latter the explicitly titled *El Sinistrado del Salonico*. Peculiarly, the British acting consul general in Salonika at the time, W.L.C. Knight, wrote that he "never heard it suggested either in the Press or in conversation with local Jews" that there was Jewish resentment of the Greek authorities for the legislation passed in the wake of the fire. Clearly he was not reading the Ladino press, which was full of such complaints. PRO/FO 608/44, W.L.C. Knight to British Secretary of State.

75. A play on the name of Venizélos's party, *El Liberal*—in Greek, *Φιλελεύθερος* (*Filelefjtheros*)—was published daily by Albert Matarasso. For a list of Salonikan Ladino papers in the early twentieth century, see Molho, *Οι Εβραίοι της Θεσσαλονίκης 1856–1919*, 127–28.

76. Most notably the Scottish historian George Finlay and David Pacifico. Both filed suit against the city of Athens when their property was confiscated by the government for the construction of King Otto's palace.

77. *La Question d'Orient vue par les Socialistes grecs: Memoire soumis par les Socialistes grecs a la Conference socialiste interalliée de Londres* (Paris, 1918).

78. Cited in "ראשית התנועה הסוציאליסטית בין יהודי שאלוניקי" [Origins of the Socialist Movement among Salonikan Jewry], in זכרון שלוניקי (Tel Aviv, 1971), 1:310.

79. The Treaty of Sèvres, signed on August 10, 1920, was negotiated between the principle allies (Britain, France, Italy, and Japan) and their nine allies, on the one side, and the Ottoman government, on the other. It was never implemented and was superseded by the Treaty of Lausanne. Fully eighteen articles of the treaty (66–83) addressed the complicated question of Smyrna. The surrounding territory was defined in article 66. Article 84 dealt with further Greek territorial gains.

80. Treat of Sèvres, article 83.

81. "Τα Νοεμβριανά των Φιλελεύθερων" [The November Riots of the Liberals], *Η Καθημερινή* [*Kathimerini*], August 13, 1920, 1. In *Η Ιστορία του Εθνικού Διχασμού* [History of the National Schism] (Athens, 1953), 245–51. (Collected writings of Eleuthérios Venizélos and Ioannis Metaxas relating to the national schism.)

82. *The Liberation of the Greek People in Turkey: An Appeal Issued by the London Committee of Unredeemed Greeks* (Manchester, UK, 1919), 3.

83. *Les Grecs en Turquie: Deux Articles, traduis de l'anglais, de la Revue THE NEW EUROPE, de Londres (Nos. des 14 et 21 novembre 1918)* (Paris, 1918), 5.

84. G. Horton, *The Blight of Asia: An Account of the Systematic Extermination of Christian Populations by Mohammedans and of the Culpability of Certain Great Powers; With a True Story of the Burning of Smyrna* (Indianapolis, 1926), 119. Cited also in Thomas W. Gallant, *Modern Greece* (London, 2001), 143.

85. Legations de Constantinople a Ministère des Affaires Etrangères (archive no. 7065, report no. 3501, June 14, 1915), cited in *Le Fait de la Semaine* 6e, no. 12 (February 9, 1918). *Les Persecutions anti-hélleniques en Turquie d'après les rapports officials des Agents diplomatiques et consulaires* (Paris, n.d.), 39–40; for

a breakdown, by location and number, of Christian deportees from Turkey in 1915, see 60–64.

86. *L'Héllenisme en Turquie Histoire Complete des Massacres: Un Plan Diabolique* (Paris, 1919), 1. "Les massacres des Grecs, organizés par les Turcs et les Allemands, comme ceux des Armeniens, avaient pour but l'estermination d'une race."

87. *Extraits de Livre "Les Secrets du Bosphore" Constantinople 1913–1916 par M. Henry Morgenthau* (Paris, n.d.), 22. Ottoman Armenians and Greeks had long seen themselves as fellow sufferers of Turkish anti-Orthodox oppression, and supported one another's national struggles. See *Greco-Armenian Agreement: Memorandum of the Oecumenical Patriarchate and of the Armenian Patriarchate of Constantinople for the Peace Conference*, Constantinople, February 11 (24), 1919 (Paris, n.d.).

88. High Commission of the French Republic, Political Service, Direction des Affaires Politiques et Commerciales, Asie Oceanie, Defrance to Pichon, no. 219, May 21, 1919, Constantinople. See also *Greek Atrocities in Asia Minor*, 2 parts (Constantinople, 1922), 3–4, 6. This pamphlet includes photographs documenting Greek atrocities, and testimonies of Greek prisoners of war regarding the actions of Greek troops and the orders they had received to commit them.

89. Eleuthérios Venizélos to the Greek House of Representatives, October 21, 1915, in Xanthaky and Sakellarios, *Greece in Her True Light*, 183.

90. See, for one example, the letter of Archbishop Germanos of Amassia and Samsoun to M. Constantinides, president of the Committee for Unredeemed Hellenes of Pont-Euxin, December 16 (29), 1918. Published in *Les Atrocités Turques au Pont-Euxin* (Paris, 1919).

91. "Η Σμύρνη Πυρπολείται και οι Κάτοικοι Σφάζονται" [Smyrna Burns and Her Residents Are Slaughtered], *Η Καθημερινή* [*Í Kathimerini*], September 2, 1922.

92. For an overview of Greece's military longings in the region, see Michael Llewlyn Smith, *Ionian Vision: Greece in Asia Minor, 1919–1922* (London, 1973).

93. For a contemporary overview, see Gustavo Traglia, *I Turchi Tornano in Europa: Dai selvaggi massacri Turchi di Smirne al tragico esodo della Tracia* (n.p., 1922). The one copy I was able to find is at Princeton University, Firestone Library, Rare Books, Pamphlets, Newsletters, and Other Ephemera on Greece and Southeastern Europe, box 1, file F, item 12.

94. The treaty was signed on July 24, 1923. For a comprehensive account, see C. Eddy, *Greece and the Greek Refugees* (London, 1931); S. P. Ladas, *The Exchange of Minorities: Bulgaria, Greece, and Turkey* (New York, 1932); A. A. Pallis, *Greece's Anatolian Venture—and After: A Survey of the Diplomatic and Political Aspects of the Greek Expedition to Asia Minor (1915–1922)* (London, 1937).

95. Cited in *Editor and Publisher* 55, no. 27, 2nd sec. (December 2, 1922): 18.

96. Venizélos Archive, Benaki Museum, Athens, 173/31, "The Refugee Problem in Greece," in *Ελληνικά Διπλωματικά Έγγραφα 1919–1940* [Greek Diplomatic Records], 3:717. To a large extent, the "exchange" was a matter of retroactive declaration. Already during the course of the last months of war and immediately afterward, hundreds of thousands of refugees had arrived in Greece.

Already as early as 1915, almost a quarter million Asia Minor Christians had been forced out of Asia Minor; see Venizélos, *Second Memorandum to King Constantine*, January 30, 1915, in Xanthaky and Sakellarios, *Greece in Her True Light*, 36. Marjorie Dobkin estimates that only 250,000 Turkish Christians went to Greece after the treaty was enacted; the rest had come already as refugees during the war and immediately after. Marjorie Housepian Dobkin, *Smyrna 1922: The Destruction of a City* (repr., Kent OH, 1988), 218.

97. See Renée Hirschon, *Heirs of the Greek Catastrophe: The Social Life of Asia Minor Refugees in Piraeus* (repr., Oxford, UK, 1998); Dimitri Pentzopoulos, *The Balkan Exchange of Minorities and Its Impact upon Greece* (The Hague, 1962).

98. "Turkey's Crimes: Hellenism in Turkey," excerpts from the *Morning Post* (Manchester), 1919, 31.

99. *Report on the Operations of the Refugees Settlement Commission for the First Three Months*, February 25, 1924, Athens.

100. *La Renaissance Juive*, December 31, 1920.

101. Cited in Georgios Yiannakopoulos, ed., *Προσφυγηκή Ελλάδα* [Refugee Greece] (Athens, 1992), 50–51.

102. Cited in Recanati, זכרון שלוניקי, 1:212.

103. *Pro-Israel*, March 21, 1919.

104. For a superb account, see Manolis Kandylakis, *Εφημεριδογραφία της Θεσσαλονίκης* [Journalism of Thessaloniki], 2 Vols. (Thessaloniki, 2000). See also Rena Molho, "*Ο Εβραϊκός τύπος της Θεσσαλονίκης*" [Salonika's Jewish Press], in *Ο Ελληνικός Εβραϊσμός* [Greek Jewish Life] (Athens, 1998), 149–69.

105. Nehama, *Histoire des Israélites*, 7:772.

106. Recanati, זכרון שלוניקי, 1:212.

107. Lausanne Convention, appendix 3, Organic Statues of the Greek Refugees Settlement Commission, article 4.

108. Cited in Recanati, זכרון שלוניקי, 1:212. It may also not have been lost on the Salonikans that Morgenthau was an opponent of Zionism.

109. Section 2, article 35.

110. *Palestine and Transjordan* 2, no. 58 (July 24, 1937): 6–7, and 2, no. 59 (July 31, 1937): 6.

111. The Balfour Declaration was a sixty-eight-word sentence addressed by A. J. Balfour to Lord Rothschild on November 2, 1917. "His majesty's government view with favor the establishment in Palestine of a national home for the Jewish people, and will use their best endeavors to facilitate the achievement of this object; it being clearly understood that nothing shall be done which may prejudice the civil and religious rights of existing non-Jewish communities in Palestine, or the rights and political status enjoyed by Jews in any other country." *Editor and Publisher* 55, no. 27, 2nd sec. (December 2, 1922): 3. To get a sense of the context for the Balfour Declaration, see *Mandate for Palestine* (Washington, DC, 1926). Department of State Division of Near Eastern Affairs, Division of Publications, Publications Series C, No. 55, Palestine, No. 1.

112. Pentzopoulos, *The Balkan Exchange of Minorities*, insert between 136 and 137.

113. William Miller, *Greece* (New York, 1928), 73.

114. Pentzopoulos, *The Balkan Exchange of Minorities*, 128.

115. *Ιστορικό Αρχείο του Υπουργείου Εξωτερικών* [Greek Foreign Ministry Archives], 1922/A/5(13), No. 3435, Eleuthérios Venizélos to Greek Ministry for Foreign Affairs, October 17, 1922, London. Cited and translated in Elisabeth Kontogiorgi, "Economic Consequences Following Refugee Settlement in Greek Macedonia, 1923–1932," in *Crossing the Aegean: An Appraisal of the 1923 Compulsory Population Exchange between Greece and Turkey*, ed. Renee Hirschon (New York, 2003), 65.

116. *Minorities in Greece: Letters Addressed to the Secretary-General of the League of Nations and the High Commissioner for Refugees, Relative to the Guarantee of the Greek Minorities Treaty of August 10, 1920* (London, 1924). See also Recanati זכרון שלוניקי, 1:212–13.

117. PRO/FO 608/44, W.L.C. Knight to British Secretary of State.

CHAPTER 6. INTERWAR GREECE: JEWS UNDER VENIZÉLOS AND METAXAS

1. Moisis Vourlas, *Έλληνας, Εβραίος, και Αριστερός* [Greek, Jew, and Leftist] (Skopelos, Greece, 2000), 7.

2. Bernard Pierron, *Juifs et Chrétiens de la Grèce Moderne* (Paris, 1976), 124. "A partir de ce moment 'être juif' prendra une signification nouvelle: cela ne voudra plus dire seulement appartenire à un groupe 'ethno-confessionnel' minoritaire et toléré parce que fort paisible, mais aussi manifester sa différence et la faire respecter dans un people qui n'était pas toujours disposé a l'accepter dans de telles conditions."

3. Law 4837/1930, adopted July 17, 1930.

4. Pierron, *Juifs et Chrétiens*, 128–29.

5. See Nicholas G. Pappas, "Concepts of Greekness: The Recorded Music of Anatolian Greeks after 1922," *Journal of Modern Greek Studies* 17 (1999): 355.

6. Elias Petropoulos, *Songs of the Greek Underworld: The Rebetika Tradition*, trans. Ed Emery (London, 2000), 75.

7. For an overview with statistical analysis, see Maria Vasilikou, "Η Εκπαίδευση των Εβραίων της Θεσσαλονίκης στο Μεσοπόλεμο" [The Education of Salonican Jews during the Interwar Period], in *Ο Ελληνικός Εβραϊσμός* [Greek Jewish Life] (Thessaloniki, 1999), 129–47.

8. Renée Hirschon, *Heirs of the Greek Catastrophe: The Social Life of Asia Minor Refugees in Piraeus* (repr., Oxford, UK, 1998), 225.

9. Minardos to Ministry of Foreign Affairs, Thessaloniki, October 26, 1929, in *Documents on the History of the Greek Jews: Records from the Historical Archives of the Ministry of Foreign Affairs*, ed. Photini Constantopoulou and Thanos Veremis (Athens, 1998), 142. I have made use of thirty of the documents included in this collection; in each case the English translation I provide is that of Constantopoulou and Veremis. The collection has drawn skeptical responses: the choice of documents, the timing of the publication, and the lack of an explicit methodology suggest that its main purpose was propagandistic. As Mark Mazower put it in a scathing essay, the reader is "offered a large quantity of materials designed to show how well Greeks have behaved toward the Jews this century and

how grateful Jews have been" (review of *The Jewish Communities of Southeastern Europe: From the Fifteenth Century to the End of World War II*; *Documents on the History of the Greek Jews: Records from the Historical Archives of the Ministry of Foreign Affairs*; and *Italian Diplomatic Documents on the History of the Holocaust in Greece [1941–1943]*, *Journal of Modern Greek Studies* 17, no. 2 [1999]: 416). The documents themselves, however, are accurately rendered and translated, as are the sources (newspaper articles, etc.) that they quote.

10. Director of Thessaloniki Press Bureau to Ministry of Foreign Affairs, July 5, 1931, Thessaloniki, in Constantopoulou and Veremis, *Documents on the History of the Greek Jews*, 175.

11. Richard Clogg, *A Concise History of Greece*, 2nd ed. (Cambridge, UK: 2002), 108–9.

12. The *idonym* law of 1929.

13. Cited in David Recanati, ed., זכרון שלוניקי: גדולתה וחורבנה של ירושלים דבלקן [Memoirs of Salonika: The Glory and Destruction of Jerusalem of the Balkans] (Tel Aviv, 1972), 1:333–34.

14. *Jewish Post*, September 6, 1936.

15. Pierron, *Juifs et Chrétiens*, 174ff.

16. Clogg, *A Concise History of Greece*, 109; Pierron, *Juifs et Chrétiens*, 175.

17. "The Case of Bulgaria vs. Her Neighbors," *Philhellene* 2, nos. 8–9 (August–September 1943): 1.

18. The Convention for the Reciprocal Emigration of Minorities, 1919, transferred all Bulgarians living in western Thrace to Bulgaria, and all Greek Orthodox Christians living in Bulgaria to Greece.

19. "The Case of Bulgaria vs. Her Neighbors."

20. Clogg, *A Concise History of Greece*, 106.

21. Recanati, זכרון שלוניקי, 1:226; PRO/FO 371/15240, no. 18 (33/3), Acting Consul General in Salonica to British Legation, June 26, 1931, Athens.

22. PRO/FO 371/15240, reg. no. C 4937/4668/19, no. 313 (278/2/31), Patrick Ramsey to the Home Office, June 26, 1931, Athens.

23. *Makedonía*, June 20, 1930.

24. PRO/FO 371/15240, no. 18 (33/3), Acting Consul General in Salonica to British Legation, June 26, 1931, Athens.

25. Among the ultranationalist organizations were Pavlo Melas, Hellas, and the National Students' Union.

26. PRO/FO 371/15240, no. 19 (33/3), Geoffrey Meade, Acting British Consul General, Salonica, to British Legation, June 30, 1931, Athens.

27. PRO/FO 371/15240, no. 313 (278/2/31), Patrick Ramsey to the British Legation, June 26, 1931, Athens.

28. Cited in Recanati, זכרון שלוניקי, 1:226–27.

29. PRO/FO 371/15240, no. 322 (278/6/31), Patrick Ramsey to British Legation, July 2, 1931, Athens.

30. PRO/FO 371/15240, reg. no. C 4937/4668/19, no. 313 (278/2/31), June 26, 1931.

31. PRO/FO 371/15240, no. 322 (278/6/31), Patrick Ramsey to British Legation, July 2, 1931, Athens.

32. PRO/FO 371/15240, no. 18 (33/3), Acting Consul General in Salonica to British Legation, June 26, 1931, Athens.

33. PRO/FO 371/15240, no. 19 (33/3), Geoffrey Meade, Acting British Consul General, Salonica, to British Legation, June 30, 1931, Athens.

34. *El Pueblo*, March 19, 1927.

35. Lucy S. Dawidowicz, *The War against the Jews, 1933–1945* (New York, 1975), 393.

36. As Yitzhak Immanuel (1899–1973) wrote in his eyewitness history of twentieth-century Salonika, "Prime Minister Metaxas, even though he declared a dictatorship, was not a hater of the Jews, and one can even say that he was somewhat supportive of them." Cited in Recanati, זכרון שלוניקי, 1:232.

37. "Αβριον θα είναι αργά," *Η Καθημερινή*, March 8, 1936. In Georgios Vlachos, ´Αρθρα στην "Καθημερινή" *(1919–1951)* [Articles in *"Kathimerini"*] (Athens, 1990), 326–27.

38. *Ριζωσπάστης [Rizospástis]*, March 1, 1936.

39. "I Kyvernisis," *Η Καθημερινή*, July 24, 1936. In Vlachos, ´Αρθρα στην "Καθημερινή," 356–57.

40. *Ελεύθερον Βήμα [Eleutheron Vyma]*, 20 and 21 April, 1936.

41. "*Η νέα κατάστασις*" [The New Situation], *Η Καθημερινή*, August 6, 1936.

42. Recanati, זכרון שלוניקי, 1:231.

43. *Μεταξάς. Λόγοι και σκέψεις 1936–1941* [Metaxas: Speeches and Ideas] (Athens, 1969), 2:10.

44. Metaxas explained the exclusion of Jews from his National Youth Organization (EON), for instance, as a provision that ensured that Jews wouldn't perceive it to be a proselytizing, Hellenizing, or Christianizing organization. "If there exist non-Christian parents, who want with all of their soul for their children to join the organization, with complete awareness that the organization is purely Christian [καθαρά Χριστιανικής], and [if] their children also want to join, we shall consider this request with goodwill, if first we can be completely sure, examining every circumstance individually, that pressure from no one—not even the slightest bit—has been exerted, neither on the parents, nor on the children." Speech, March 5, 1939, in *Μεταξάς. Λόγοι και σκέψεις*, 2:11. Many Jews did in fact try to join EON, but whether or not they were accepted depended on chapter leaders and local demographics. In regions where few Jews lived, the chances of approval were greater. In heavily Jewish areas (notably Salonica), Jews were allowed only to participate in parallel organizations. And even those who were admitted weren't fully assimilated into the group; for public events they were encouraged to wear their school uniforms, rather than those of EON. *Πλεκτά* [Knitted], publication on EON membership attire, "Pamphlets, Newsletters, and Other Ephemera on Greece and Southeastern Europe," box 2, file A, item 7, Firestone Library, Special Collections, Princeton University.

45. Recanati, זכרון שלוניקי, 1:487.

46. Ibid., 1:232.

47. "Ολοί λοιπόν οι Έλληνες είναι ορθόδοξοι" October 19, 1939, in *Μεταξάς. Το προσωπικό του ημερολόγιο 1933–1941* [Metaxas: His Personal Diaries, 1933–1941] (Athens, 1987), 8:770.

48. Cited in Joachim G. Joachim, *Ioannis Metaxas: The Formative Years, 1871–1922* (Mannheim, Ger., 2000), 15–16.

49. Cited in ibid., 41–42.

50. March 25, 1938, in *Μεταξάς. Το προσωπικό του ημερολόγιο 1933–1941*, 7:299.

51. Neville Chamberlain to the House of Commons, April 13, 1939, in *Diplomatic Documents: Italy's Aggression against Greece* (Athens, 1940), no. 31:20–21; M. Daladier to the Press, April 13, 1939, in *Diplomatic Documents*, no. 33:21.

52. January 30, 1938, in *Μεταξάς. Το προσωπικό του ημερολόγιο 1933–1941*, 7:293.

53. See, for example, January 9, 1938, February 19, 1938, March 22, 1938, in *Μεταξάς. Το προσωπικό του ημερολόγιο 1933–1941*, 7:292, 293, 299. On March 22, 1938, he wrote, "Sometimes I see my life all black."

54. April 10, 1938, in *Μεταξάς. Το προσωπικό του ημερολόγιο 1933–1941*, 7:300.

55. September 23, 1928, Rome. See *Diplomatic Documents*, no. 1:7–8.

56. Cited in *Diplomatic Documents*, no. 5:11.

57. Mavroudis (Greek Under-Secretary of State for Foreign Affairs) to Royal Legation in Tirana, April 5, 1939, in *Diplomatic Documents*, no. 11:13.

58. Simopoulos to Royal Ministry for Foreign Affairs, April 6, 1939, in *Diplomatic Documents*, no. 13:14.

59. Skeferis to Royal Ministry for Foreign Affairs, April 7, 1939, in *Diplomatic Documents*, no. 14:14.

60. Metaxas to the Minister in Rome, April 8, 1939, in *Diplomatic Documents*, no. 19:15.

61. Memorandum, April 9, 1939, in *Diplomatic Documents*, no. 21:16.

62. Ioannis Metaxas to the Royal Legation in Rome, April 10, 1939, *Diplomatic Documents*, no. 27:19.

63. February 18, 1940, March 3, 1940, and March 17, 1940, in *Μεταξάς. Το προσωπικό του ημερολόγιο 1933–1941*, 8:453, 455, 457.

64. August 15, 1940, *Μεταξάς. Το προσωπικό του ημεπολόγιο 1933–1941*, 8:495.

65. *Η Καθημεπινή*, August 17, 1940.

66. Stanley Casson, *Greece against the Axis* (Washington, DC, 1943), 12.

67. Benito Mussolini to Ioannis Metaxas, October 28, 1940, in *Diplomatic Documents*, no. 178:133–34.

68. Metaxas's Proclamation to the Greek People, October 28, 1940, in *Diplomatic Documents*, no. 181:136.

69. *Lifeline: Organ of the American Jewish Joint Distribution Committee* 1, nos. 1–2 (February 1941): 6.

70. *United Palestine Appeal 1941 Yearbook* (New York, 1941), 22, 24. Between 1933 and 1940, 280,000 Jews immigrated to Israel; less than 3,000 were Greek.

71. Ostensibly there were approximately thirteen thousand Jews serving in the Greek army on the eve of the German occupation. In Errikos Sevillias, *From Athens*

to Auschwitz, trans. and intro. Nikos Stavroulakis (Athens, 1993), 93n1. See also Dawidowicz, *The War against the Jews,* 393. These numbers seem high, however.

72. Frizis had entered the officer corps before the Metaxas regime broke with tradition and barred Jews entry.

73. Introduction to Constantopoulou and Veremis, *Documents on the History of the Greek Jews,* 34.

74. Testimony of Itzchak Nechama, Eichmann trial session 47, in *The Trial of Adolf Eichmann* (Jerusalem, 1992), 2:850.

75. Casson, *Greece against the Axis,* 19–20.

76. *Οι μακαρονάδες* [The Macaroni Eaters] (Athens, n.d.). Collection of cartoons and satire mocking fascist Italy, "Pamphlets, Newsletters, and Other Ephemera on Greece and Southeastern Europe," box 3, file C, item 3, Firestone Library, Special Collections; Princeton University. *Ο Μουσολίνι η αρκούδα* [Mussolini: The Bear] (Athens, n.d.). "Pamphlets, Newsletters, and Other Ephemera on Greece and Southeastern Europe," box 3, file C, item 4, Firestone Library, Special Collections, Princeton University.

77. Appendix, Zannini to the "Ferrara" (P.M. 52-A, October 26, 1940/XVIII), in *Diplomatic Documents,* 139–40. ("Da diciannove mesi in questa forte e rude terra de Albania, tempriamo armi e cuori, tesi ad una meta ormai vicina.")

78. Transcription of Italian cabinet meeting of October 15, 1940, reproduced in *Philhellene* 3, nos. 9–10 (September–October 1944): 8–11.

79. See Alexander Papagos, *The German Attack on Greece* (London, 1946). Papagos was the commander in chief of the Greek army during the Greco-Italian and Greco-German campaigns, 1940–41.

80. Casson, *Greece against the Axis,* 97.

CHAPTER 7. OCCUPATION AND DEPORTATION: 1941–44

1. Nikos Stavroulakis, *Faces and Facets: The Jews of Greece. Photographs by Morrie Camhi* (New Rochelle, NY, 1995), 104–5.

2. 260 Jews from Hania and 5 families from Rethymnon. *Χρονικά,* April 1984. See the testimony of Nikos Sgourakis in Michael Matsas, *The Illusion of Safety: The Story of the Greek Jews during the Second World War* (New York, 1997), 157–58. Also see Nikolaos Zevgadaki, "Οι Εβραίοι της Κρήτης κατά την Γερμανικήν κατοχήν" [The Jews of Crete during the German Occupation], *Χρονικά,* January–February, 2001.

3. On Crete, most Jews had names that were indistinguishable from Christians', and it is likely that several of them were shot by Nazis in the general Nazi reprisals against the local population, without their murderers necessarily having known that they were Jews. There is a recorded oral account of two Jewish Cretan families who survived the war hidden in the home of Anastasios Karvoulakis in Iraklion, in G. C. Kiriakopoulos, *The Nazi Occupation of Crete, 1941–1945* (Westport, CT, 1995), 35–37. See also *Χρονικά,* April 1984, according to which seven Cretan Jews survived the occupation in hiding. A Haniote Jew, Alberto Minervo, who worked for the British secret service during the war, escaped to Athens in spring 1943. See Matsas, *The Illusion of Safety,* 159–61.

4. Katina Sygkellaki, "Μέσα σε ένα εικοσιτετράωρο" [Within Twenty-four Hours], *Εβραϊκή Εστία* [*Evraiki Estia*], July 1, 1947. Cited also in Fragkiski Ampatzopoulou, ed., *Το Ολοκαύτωμα στης Μαρτυρίες των Ελλήνων Εβραίων* [The Holocaust in Greek Jewish Testimonies] (Thessaloniki, 1993), 205–7.

5. For a remarkable account of the saving of the Zákinthos Jews, see Dionysis Stravolemos, *Heroism—A Justification: The Saving of the Jews of Zakynthos during the Occupation*. Marcia Haddad Ikonomopoulos, unpublished translation, 25. Courtesy Marcia Haddad Ikonomopoulos and the Kehila Kedosha Janina, New York. See also Dion. Stravolemos and Ntinou Konomos, "Γύρω από τη διάσωση των Εβραίων της Ζακύνθου κατά τα χρόνια της κατοχής" [On the Saving of the Jews of Zakynthos during the Years of the Occupation]. *Χρονικά*, April–May 1996; *Η Καθημερινή*, November 8, 1978.

6. "If you carry out the deportations, I will go with the Jews and I will share their fate." *Χρονικά*, January 1984; *Ta Nea Mas*, June 1, 1992.

7. Michael Molho, *In Memoriam: Hommage aux Victimes Juives des Nazis en Grèce* (Thessaloniki, 1988), 229–30.

8. Stravolemos, *Heroism*, p. 25. The effect of the earthquakes was likened to the bombing of Hiroshima. *Ionian Tragedy: The Greek Earthquakes Aug. 9–12, 1953* (n.d.).

9. Moshe Beoski, President of the Committee of the Righteous to Mrs. V. Stavroulomenos [*sic*] (Surviving Sister of Archbishop Hrysostomos), July 31, 1978, Jerusalem. Courtesy of the Kehila Kedosha Janina. Hrysostomos died in 1958; Karrer attended the ceremony.

10. *Τα Νέα Μας* [*Ta Néa Mas*], June 1, 1992; *Χρονικά*, September–October, 1992.

11. Stravolemos, *Heroism*, 3.

12. Speech of Samuel Mordos, June 14, 1992, Zákinthos. Courtesy of the Kehila Kedosha Janina.

13. See Mark Mazower, *Inside Hitler's Greece: The Experience of Occupation, 1941–44* (New Haven, CT, 2001), 253.

14. The community of Rhodes was deported later. Rhodes, along with the rest of the Dodecanese, did not become part of Greece until several years after the war.

15. A copy is in the Jewish Museum of Athens.

16. Bracha Rivlin, יון :הקהילות פּנקס [Encyclopedia of Jewish Communities: Greece] (Jerusalem, 1998), 368.

17. *Jewish Museum of Greece Newsletter* 10 (September 1983): 1.

18. Steven Bowman, *Jewish Resistance in Wartime Greece* (London, 2006). See also Matsas, *The Illusion of Safety*, which includes memoirs of Greek Jewish resistance fighters (269–326).

19. Molho, *In Memoriam*, 53; Mazower, *Inside Hitler's Greece*, 237–38.

20. Testimony of Itzchak Nechama, Eichmann trial session 47, in *The Trial of Adolf Eichmann* (Jerusalem, 1992), 2:850.

21. Giomtov Giakoel, *Απομνημονεύματα 1941–1943* (Thessaloniki, 1993), 49.

22. Cited in Rebecca Fromer, *The House by the Sea: A Portrait of the Holocaust in Greece* (San Francisco, 1997), 60–61.

23. Zamboni to Italian Diplomatic Mission to Athens, July 30, 1942, Salonica, in *Italian Diplomatic Documents on the History of the Holocaust in Greece (1941–1943)* (Tel Aviv, 1999), 1942.12, 97–98. The consulates of Romania, Turkey, France, Hungary, Denmark, Finland, Sweden, Portugal, Spain, and Switzerland were closed.

24. PRO/FO 371/51171 98605, 161/21/45, Refugee Department, London, to British Embassy, March 13, 1945, Madrid.

25. Pietro Vitelleschi to Italian Diplomatic Mission to Athens, October 30, 1941, Salonica, in *Italian Diplomatic Documents*, 1941.4, 72

26. "In generale essi non grandiscono il nostro intervento in favore dei cittadini greci." Zamboni to Italian Diplomatic Mission to Athens, July 23, 1942, Salonika, in *Italian Diplomatic Documents*, 1942.8, 89.

27. Ελευθερία [*Eleftheria*], September 10, 1942.

28. Ibid., October 18, 1942.

29. Stephen G. Xidis, *The Economy and Finances of Greece under Occupation* (New York, n.d.), 12.

30. Ελευθερία, October 28, 1942; Bulletin of the Italian General Consulate, January 5, 1943, Salonika, in *Italian Diplomatic Documents*, 1943.1, 119.

31. Stanley Casson, *Greece against the Axis* (Washington, DC, 1943), 101.

32. Cited in frontispiece in Casson, *Greece against the Axis*.

33. *New Judaea* 19, no. 3 (December 1942): 39–40.

34. State Attorney Bar-Or, on Document No. 997, Suhr to Rademacher (German Foreign Ministry), July 11, 1942, Eichmann trial session 47, in *The Trial of Adolf Eichmann*, 2:848.

35. Violetta Hionidou, *Famine and Death in Occupied Greece, 1941–1944* (New York, 2006), 35–38. This is a superb study of the famine produced by the occupation and the politics of food in occupied Greece.

36. Giakovos Hantali, Από το Λευκό Πύργο στης Πύλες του 'Αουσβιτς [From the White Tower to the Gates of Auschwitz] (Thessaloniki, 1996), 53–54.

37. The order was published in *Das neue Europa* (Νέα Ευρώπη) [*New Europe/Néa Evrópi*]. Document No. 117 (exhibit T/974), Eichmann trial session 47, in *The Trial of Adolf Eichmann*, 2:850.

38. Molho, *In Memoriam*, 59. See also Yiakoel, Απομνημονεύματα, 57–58. See also Testimony of Itzchak Nechama, Eichmann trial session 47, in *The Trial of Adolf Eichmann*, 2:852.

39. Testimony of Itzchak Nechama, Eichmann trial session 47, in *The Trial of Adolf Eichmann*, 2:853.

40. Burton Berry, U.S. consul in Istanbul, dispatches to Washington. Cited in Matsas, *The Illusion of Safety*, 34–35.

41. Yiakoel and Ampatzopoulou, Απομνημονεύματα, 58.

42. Νέα Ευρώπη [*Néa Evrópi*], July 12, 1942.

43. Michael Molho, *In Memoriam: Αφιέρωμα εις Μνήμην των Ισραηλίτων θυμάτων του ναζισμού εν Ελλάδι [An Offering to the Memory of the Jewish Victims of Nazism in Greece] (Thessaloniki, 1976), 82.

44. In זכרון שלוניקי, 1:238.

45. *Ελευθερία*, May 11, 1942. The *Εθνικός Απελευθερωτικός Μέτωπος* (EAM, or the National Liberation Front) was one of the two main resistance movements of the occupation period.

46. Consul General Zamboni to Ministry for Foreign Affairs, Rome, July 9, 1942, Salonika, in *Italian Diplomatic Documents*, 1942.5, 80–81; Zamboni to Italian Diplomatic Mission in Athens, July 16, 1942, Salonika, in *Italian Diplomatic Documents*, 1942.7, 83–84.

47. Testimony of Itzchak Nechama, Eichmann trial session 47, in *The Trial of Adolf Eichmann*, 2:854.

48. Bulletin of the Italian General Consulate, November 8, 1942, Salonika, in *Italian Diplomatic Documents*, 1942.24, 116.

49. Testimony of Palompas Allalouf, in Erika Kounio Amariglio and Almpertos Nar, *Προφορικές Μαρτυρίες Εβραίων της Θεσσαλονίκης για το Ολοκαύτωμα* [Salonikan Jewish Testimonies on the Holocaust] (Thessaloniki, 1998), 33. One of Allalouf's brothers died in forced labor when a German guard struck him on the head. Allalouf's number in Auschwitz was 39.239.

50. Yiakoel and Ampatzopoulou, *Απομνημονεύματα*, 59.

51. Reports on this are unclear and contradictory. Many members of the Jewish community later held Koretz culpable for the destruction of the cemetery and regarded him as a collaborator with the Germans. He was named as such at the Adolf Eichmann trial. In terms of handing over the cemetery, though, it seems that he had little choice in the matter. Moreover, the Greek Christians of the city had for some time been agitating for the vast piece of land occupied by the cemetery. The Germans recognized it as a "wedge issue" with which they could further exacerbate the antagonism between Jews and Christians.

52. Report of P. Kontopoulos, September 15, 1943, Cairo. Cited in Photini Constantopoulou and Thanos Veremis, eds., *Documents on the History of the Greek Jews: Records from the Historical Archives of the Ministry of Foreign Affairs* (Athens, 1998), 257. Koretz's compliance with demand that he turn the community's documents over to the German authorities greatly facilitated the rapid liquidation of the Jewish community by the SS. He is remembered by many as a traitor and collaborator.

53. Bulletin of the Italian General Consulate, November 8, 1942, Salonika, in *Italian Diplomatic Documents*, 1942.24, 116.

54. Constantopoulou and Veremis, *Documents on the History of the Greek Jews*, 274. See also Testimony of Itzchak Nechama, Eichmann trial session 47, in *The Trial of Adolf Eichmann*, 2:854. Nechama gives the ransom figure as 2.5 billion drachmas.

55. See *Italian Diplomatic Documents*, 84n2.

56. A. L. Molho, Notes on the Present Situation of Greek Jewry, October 12, 1943, Cairo, in Constantopoulou and Veremis, *Documents on the History of the Greek Jews*, 282.

57. Testimony of Palompas Allalouf, in Kounio Amariglio and Nar, *Προφορικές Μαρτυρίες Εβραίων της Θεσσαλονίκης για το Ολοκαύτωμα*, 33.

58. Testimony of Oro Alfantari, in Kounio Amariglio and Nar, *Προφορικές Μαρτυρίες Εβραίων της Θεσσαλονίκης για το Ολοκαύτωμα*, 45.

59. Zamboni to Diplomatic Mission to Athens, July 23, 1942, Salonika, in *Italian Diplomatic Documents*, 1942.9, 90–94; Ghigi to Consul General in Salonika, August 1, 1942, Athens, in *Italian Diplomatic Documents*, 1942.13, 99–100; Zamboni to Italian Diplomatic Mission to Athens, August 13, 1942, Salonika, in *Italian Diplomatic Documents*, 1942.16, 104–6.

60. "Sarà opportune che codesto Ufficio sospenda iniziativa da esso presa in favore ebrei di cittadinanza ellenica." D'Ajeta to Italian Embassy in Berlin, August 14, 1942, Rome, in *Italian Diplomatic Documents*, 1942.18, 110.

61. In זכרון שלוניקי, 1:238–40.

62. Yiakoel and Ampatzopoulou, *Απομνημονεύματα*, 86.

63. Molho, *In Memoriam*, 75.

64. Eichmann trial session 47, in *The Trial of Adolf Eichmann*, 2:849–50.

65. "e severamente proibito entrare, per qualsiasi motivo, nei quartieri ebrei. Si fa presente che I militari tedeschi di guardia hanno l'ordine di sparare." Italian Military Base 40, Salonika, in *Italian Diplomatic Documents*, 1943.9, 128.

66. Erika Kounio Amariglio, *From Thessakoniki to Auschwitz and Back: Memories of a Survivor from Thessaloniki* (London, 2000), 43.

67. Zamboni to Italian Diplomatic Mission to Athens, Saul Moisis to Zamboni, February 17, 1943, Salonika, in *Italian Diplomatic Documents*, 1943.5, 125–27. Moisis was a legal consultant to the Italian Consulate in Salonika. See also the various correspondence of Dieter Wisliceny and Max Merten to the Jewish Community, February 6–13, 1943, in Molho, *In Memoriam*, 76–77.

68. Testimony of Itzchak Nechama, Eichmann trial session 47, in *The Trial of Adolf Eichmann*, 2:850–51.

69. Ghigi to Ministry for Foreign Affairs, Rome, May 19, 1942, Athens, in *Italian Diplomatic Documents*, 1942.2, 76–79; d'Ajeta to Italian Diplomatic Mission to Athens, May 30, 1942, Rome, in *Italian Diplomatic Documents*, 1942.3, 76–79; Bulletin of Italian General Consulate, July 8, 1942, Salonika, in *Italian Diplomatic Documents*, 1942.4, 76–79.

70. State Attorney Bar-Or, on Document No. 1003, Schonberg to the Foreign Ministry (February 26, 1943), Eichmann trial session 47, in *The Trial of Adolf Eichmann*, 2:850.

71. Amariglio, *From Thessakoniki to Auschwitz*, 42.

72. Fromer, *The House by the Sea*, 41.

73. Hantali, *Από το Λευκό Πύργο*, 71.

74. Information from Salonika, July 10, 1943, in Constantopoulou and Veremis, *Documents on the History of the Greek Jews*, 270.

75. Jews of Italian Citizenship to Zamboni, March 8–9, 1943, Salonika, in *Italian Diplomatic Documents*, 1943.14, 134–35; Ghigi to Pietrocarchi, Athens, in *Italian Diplomatic Documents*, 1943.46, 179–80.

76. Zamboni to Italian Diplomatic Mission to Athens, March 6, 1943, Salonika, in *Italian Diplomatic Documents*, 1943.12, 131–32.

77. Zamboni to Italian Diplomatic Mission to Athens, March 14, 1943, Salonika, in *Italian Diplomatic Documents*, 1943.15, 137.

78. The first left Salonika on March 15, 1943, and the last on August 10. *Μαύρη βίβλος της κατοχής*/Schwarzbuch der Besatzung [Black Book of the Occupation] (Athens, 1998), 51.

79. Zamboni to Italian Diplomatic Mission to Athens, March 18, 1943, Salonika, in *Italian Diplomatic Documents*, 1943.16, 138.

80. Testimony of Menachem Stroumsa, excerpted in Ampatzopoulou, *To Ολοκαύτωμα στης Μαρτυρίες των Ελλήνων Εβραίων*, 177–78.

81. Testimony of Rabbi Azaria of Veroia, excerpted in Ampatzopoulou, *To Ολοκαύτωμα στης Μαρτυρίες των Ελλήνων Εβραίων*, 180–81.

82. *New Judaea* 19, no. 9 (June 1943): 135.

83. *New Judaea*, 20, no. 4 (January 1944): 60.

84. "Circolano insistentemente in questi giorni voci, non si sa con quale fondamento, di prossime misure contro gli ebrei di Salonicco. Si parla di deportazioni in Polonia. . . . Naturalmente cio ha messo in agitazione questa grossa Comunita israelitica." Zamboni to Italian Diplomatic Mission to Athens, January 23, 1943, Salonika, in *Italian Diplomatic Documents*, 1943.3, 121.

85. *Νέα Ευρώπη*, February 27, 1943.

86. Zamboni to Italian Diplomatic Mission to Athens, February 28, 1943, Salonika, in *Italian Diplomatic Documents*, 1943.11, 130.

87. Giorgos Ioannou, *Το δικό μας αίμα* [Our Own Blood] (Athens, 1978), 60–61. Also cited in Mazower, *Inside Hitler's Greece*, 245–46.

88. Markos Nahon, *Μπίρκεναου το Στρατόπεδον του Θανάτου* [Birkenau, the Death Camp], trans. Asser Moissis (Thessaloniki, 1991), 36.

89. *Jewish Chronicle*, December 3, 1943.

90. Members of the various resistance groups seem to have had better knowledge, earlier, of what was being done to Europe's Jews than others in Greece did. Yosef Barzilai spent the war in hiding with Greek Christians. In a later testimony he commented, "It is interesting to note that both rival camps during the civil war were considerate toward Jews, apparently because they already knew what was happening to Jews in Europe, so my movements in Athens were not restricted." See interview with Yosef Barzilai in Shmuel Refael, ed., בנתיבי שאול: יהודי יוון בשואה [Road to Hell: Greek Jews in the Holocaust] (Jerusalem, 1988), 121.

91. Markos Vafiadis, *Απομνημονεύματα* [Memoirs] (Athens, 1985), 2:87.

92. Zamboni to Italian Diplomatic Mission to Athens, March 20, 1943, Salonika, in *Italian Diplomatic Documents*, 1943.17, 139–40.

93. A. L. Molho, Notes on the Present Situation of Greek Jewry, October 12, 1943, Cairo, in Constantopoulou and Veremis, *Documents on the History of the Greek Jews*, 259.

94. Zamboni to Italian Diplomatic Mission to Athens, March 27, 1943, Salonika, in *Italian Diplomatic Documents*, 1943.20, 145.

95. Pieche to Ministry for Foreign Affairs, March 23, 1943, Rome, in *Italian Diplomatic Documents*, 1943.19, 142–44. The Greek Jews deported from the Bulgarian zone were almost all killed in Treblinka.

96. Zamboni to Italian Diplomatic Mission to Athens, April 22, 1943, Salonika, in *Italian Diplomatic Documents*, 1943.38, 171.

97. Zamboni to Italian Diplomatic Mission to Athens, the head of the Jewish police was the feared and despised Vital Hasson, to whom the Germans had given great power, "which he used with singular ferocity against his coreligionists." In the end, he too was targeted for deportation, and the Italians worked hard to prevent him from entering the Italian zone, where they feared he would help the

Germans track down Jews. Hasson escaped with his family, but was arrested by the British and turned over to the Greek authorities after the war. He was sentenced to death and executed on Corfu in March 1948. August 11, 1943, Salonika, in *Italian Diplomatic Documents*, 1943.119, 193–94, 292–95, 294–95n16.

98. Fromer, *The House by the Sea*, 100.

99. Zamboni to Italian Diplomatic Mission to Athens, April 27, 1943, Salonika, in *Italian Diplomatic Documents*, 1943.43, 175–76.

100. Zamboni to Italian Diplomatic Mission to Athens, May 1, 1943, Salonika, in *Italian Diplomatic Documents*, 1943.47, 181–82.

101. Zamboni to Italian Diplomatic Mission to Athens, May 1, 1943, Salonika, in *Italian Diplomatic Documents*, 1943.48, 182–83; Bastianini to the Italian Embassy in Berlin, May 1, 1943, Salonika, in *Italian Diplomatic Documents*, 1943.48, 190–93.

102. Zamboni to Italian Diplomatic Mission to Athens, May 13, 1943, Salonika, in *Italian Diplomatic Documents*, 1943.56, 194–95.

103. September 30, 1943, Thessaloniki, in Constantopoulou and Veremis, *Documents on the History of the Greek Jews*, 274.

104. Castruccio to Italian Diplomatic Mission to Athens, July 15, 1943, Salonika, in *Italian Diplomatic Documents*, 1943.102, 257–69. There are discrepancies in the figure: lists written up by the Italian consulate on the train's departure include 331 names; when they were met in Athens, there were reportedly 350 people aboard. Ghigi to Ministry for Foreign Affairs, August 13, 1943, Athens, in *Italian Diplomatic Documents*, 1943.121.

105. Castruccio to Italian Diplomatic Mission to Athens, July 24, 1943, Salonika, in *Italian Diplomatic Documents*, 1943.107, 277.

106. Castruccio to Ministry for Foreign Affairs, August 11, 1943, Salonika, in *Italian Diplomatic Documents*, 1943.120, 296–97. See also State Attorney Bar-Or on exhibit T/988, Eichmann trial session 47, in *The Trial of Adolf Eichmann*, 2:857.

107. "La colonia ebrea di Salonicco, che era stata fondata prima della scoperta dell'America e che contava circa 60.000 persone non esiste piu. . . . La liquidazione della colonia israelita si e svolta e si e consumata in mezzo ad atrocita, orrori e delitti, come non avevo sentito raccontare nella storia di tutti i tiempi e di tutti i popoli." Castruccio to Ministry for Foreign Affairs, August 11, 1943, Salonika, in *Italian Diplomatic Documents*, 1943.120, 296–97.

108. Railroad records indicate that 42,830 Jews were sent to Aushwitz-Birkenau from Salonika alone, and another 3,000 from the Bulgarian zone. Ioannou, *Το δικό μας αίμα*, 61. The 38,386 figure is cited in Mazower, *Inside Hitler's Greece*, 244.

109. Rivlin, פנקס הקהילות, 194.

110. Xidis, *The Economy and Finances of Greece*, 13–15.

111. Testimony of Mois Pessach, excerpted in Ampatzopoulou, *Το Ολοκαύτωμα στης Μαρτυρίες των Ελλήνων Εβραίων*, 186.

112. *Spectator* (London), July 30, 1943.

113. *Philhellene* 2, nos. 8–9 (August–September 1943): 2.

114. *Jewish Chronicle*, December 3, 1943.

115. Testimony of Pavlos Simcha, excerpted in Ampatzopoulou, *To Ολοκαύτωμα στης Μαρτυρίες των Ελλήνων Εβραίων*, 164. Simcha's family was from Kavála and left for Athens in 1940.

116. Eichmann trial session 47, in *The Trial of Adolf Eichmann*, 2:842–46.

117. Pieche to Ministry for Foreign Affairs, April 1, 1943, Rome, in *Italian Diplomatic Documents*, 1943.22, 148–50.

118. A. L. Molho, Notes on the Present Situation of Greek Jewry, October 12, 1943, Cairo, in *Documents on the History of the Greek Jews*, 283.

119. Testimony of Sol Kazes, in Kounio Amariglio and Nar, *Προφορικές Μαρτυρίες Εβραίων της Θεσσαλονίκης για το Ολοκαύτωμα*, 74–75.

120. Testimony of Maurice Benveniste, cited in Matsas, *The Illusion of Safety*, 170. Also excerpted in Ampatzopoulou, *To Ολοκαύτωμα στης Μαρτυρίες των Ελλήνων Εβραίων*, 181–84.

121. Matsas, *The Illusion of Safety*, 76.

122. Gioa Koen, in Refael, נתיבי שאול, 238–40.

123. Michael Molho and Josef Nehama, שואת יהודי יוון [The Destruction of Greek Jewry] (Jerusalem, 1965): 98.

124. Testimony of Evangelitsa Hamouri (1959), excerpted in Ampatzopoulou, *To Ολοκαύτωμα στης Μαρτυρίες των Ελλήνων Εβραίων*, 184–85.

125. Testimony of Mois Pessach (1959), excerpted in Ampatzopoulou, *To Ολοκαύτωμα στης Μαρτυρίες των Ελλήνων Εβραίων*, 186.

126. Testimony of Yuda Perahia (1959), excerpted in Ampatzopoulou, *To Ολοκαύτωμα στης Μαρτυρίες των Ελλήνων Εβραίων*, 188. Perahia was the last Jew living in Xanthi in 1959.

127. May 1943, Didimothicho, in Constantopoulou and Veremis, *Documents on the History of the Jews of Greece*, 271–72. The documents indicate, mistakenly, that the roundup occurred on May 4, not March 4.

128. Thrasyvoulos Papastratis, *Οι Εβραίοι του Διδυμότειχου* [The Jews of Didimoticho] (Athens, 2001), 61–62.

129. Yaakov Jabari, in Refael, נתיבי שאול, 132–34.

130. *New Judaea* 20, no. 4 (January 1944): 60.

131. Molho, *In Memoriam*, 185

132. *New Judaea* 19, no. 9 (June 1943): 135.

133. For the testimonial of a Jewish family in hiding in Athens with a Christian family during the occupation, see Lilian Mpenroumpi-Ampastado, *Τα Τετράδια της Λίνας. Ενα Ντοκουμέντο απο την Κατοχή* [Lina's Notebooks: A Document from the Occupation] (Thessaloniki, 1999).

134. Marsel Natzari, *Χρονικό 1941–1945* [Chronicle] (Thessaloniki, 1991), 21. Written account found between crematoriums I and II at Birkenau, October 24, 1980. Natzari survived the war, and went on to write an account of his years in hiding with the Greek resistance, capture by the SS, and time in Auschwitz-Birkenau. A jokester with a wonderful sense of humor, Natzari (Nadjari) was nicknamed "Marcel, *il loco*." Recruited for the Sonderkommando in Birkenau, "He cheered the prisoners and made one or two of the Germans laugh at his antics—and this was extraordinary." Fromer, *A House by the Sea*, 43.

135. The Jews of Preveza made up one of the Sonderkommando teams that staffed the crematoriums in Auschwitz. See Molho, *In Memoriam*, 214. Given the letter's location, it was likely written by a member of the Sonderkommando.

136. Reuven Dafni and Yehudit Kleiman, eds., *Final Letters* (London, 1991), 122. Cited also in Mazower, *Inside Hitler's Greece*, 257–58. "Προς τους αγαπημέ νους μου . . . η αγαπημένη μου παρέα . . . και . . . προς την αγαπημένη μου πατρίδα 'ΕΛΛΑΣ.'")

137. על יוון, מולדתנו שמלפני המלחמה, interview with Shaul Chazan, in Gideon Greif, ed., בכינו בלי דמעות: עדויותיהם של אנשי הזונדרקומאנדו היהודים מאושוויץ [We Wept without Tears: Testimonies of the Jewish Sonderkommando from Auschwitz] (Tel Aviv, 1999), 312.

138. Eichmann trial session 47, in *The Trial of Adolf Eichmann*, 2:858.

139. Errikos Sevillias, *Athens-Auschwitz*, trans. Nikos Stavrouakis (Athens, 1983), 3.

140. See Alexander Kitroeff, *War-Time Jews: The Case of Athens* (Athens, 1995), 55.

141. Fromer, *The House by the Sea*, 100; Amariglio, *From Thessaloniki to Auschwitz*, 48.

142. A complete chronology of events during the German occupation is found in *Μαύρη βίβλος της κατοχής*/Schwarzbuch der Besatzung.

143. *New York Times*, March 1, 1942.

144. "News from Greece: Tyranny, Hunger, Despair," *Philhellene* 1, no. 6 (June 1942): 2.

145. Polopoulos et al. to Gotzamanis, March 20, 1943, Athens; Damaskinos et al. to Logothetopoulos, March 23, 1943, Athens; Damaskinos et al. to von Altenburg, March 24, 1943, Athens. All in Constantopoulou and Veremis, *Documents on the History of the Greek Jews*, 249–56.

146. *New Judaea* 21, nos. 11–12 (August–September 1945): 180–81.

147. Information from Athens, September 9, 1943, in Constantopoulou and Veremis, *Documents on the History of the Greek Jews*, 272.

148. Xidis, *The Economy and Finances of Greece*, 42–45.

149. Cited in Mazower, *Inside Hitler's Greece*, 251.

150. Report of a Refugee from Athens, October 1943, Istanbul, in Constantopoulou and Veremis, *Documents on the History of the Greek Jews*, 284–86.

151. Fromer, *The House by the Sea*, 82–90.

152. הארץ [*Ha-Aretz*], April 13, 1944.

153. A. L. Molho, Notes on the Present Situation of Greek Jewry, October 12, 1943, Cairo, in Constantopoulou and Veremis, *Documents on the History of the Greek Jews*, 261.

154. Kostas Triantafillidis , cited in *Η Καθημερινή*, May 10, 1976. Cited also in Matsas, *The Illusion of Safety*, 89.

155. *New Judaea* 21, nos. 1–2 (October–November 1944): 38.

156. *New Judaea* 20, no. 4 (January 1944): 60.

157. See, for example, Prato to Ghigi, April 19, 1943, Athens, in *Italian Diplomatic Documents*, 1943.35, 167–68.

158. Molho, *In Memoriam*, 185. For a complete summary account of Barzilai's dealings with Wisliceny, see ibid., 185–92.

159. Prato to Ghigi, April 24, 1943, Athens, in *Italian Diplomatic Documents*, 1943.41, 173.

160. Report of a Refugee from Athens, October 1943, Istanbul, in Constantopoulou and Veremis, *Documents on the History of the Greek Jews*, 284–86; *Ellas*, November 26, 1943, London, in Constantopoulou and Veremis, *Documents on the History of the Greek Jews*, 286–87. For a summary account, see Mazower, *Inside Hitler's Greece*, 250–51.

161. *Ελεύθερη Ελλάδα* [*Eléftheri Elláda*], October 14, 1943. Cited in Kitroeff, *War-Time Jews*, 63.

162. For one testimony of a Greek Jew who hid with the partisans, see anonymous, "Οι αντάρτες σώζουν Εβραίους. Από το ημερολόγιο ενός Εβραίου αντάρτη" [The Andartes Save Jews: From the Diary of a Jewish Andarti], *Ισραηλιτικόν Βήμα* [*Israilitikün Víma*], May 17, 1946. Excerpted in Ampatzopoulou, *To Ολοκαύτωμα στης Μαρτυρίες των Ελλήνων Εβραίων*, 112–15.

163. Eliyahu Barzilai, cited in פרקי זכרונות :יהדות יוון בחורבנה [Greek Jewry in the Holocaust: Memoirs] (Tel Aviv, 1988), 25.

164. *Jewish Chronicle*, December 3, 1943; Report of a Refugee from Athens, October 1943, Istanbul, in Constantopoulou and Veremis, in *Documents on the History of the Greek Jews*, 288–89, 285; Molho, *In Memoriam*, 194.

165. See, for example, Ephraim Oshry, *Responsa from the Holocaust* (New York, 1983), 64.

166. Molho, *In Memoriam*, 196–97.

167. הארץ, May 7, 1944.

168. Report of a Refugee from Athens, October 1943, Istanbul, in Constantopoulou and Veremis, *Documents on the History of the Greek Jews*, 285; see also Executive Committee of the World Jewish Congress to Ouziel, April 28, 1944, in אוצר יהודי ספרד [Treasures of the Jews of Sefarad] (Jerusalem, 1959), 2:165.

169. "Homeless in Greece," *Philhellene* 3, nos. 7–8 (July–August 1944): 6.

170. הארץ, May 7, 1944.

171. הבוקר [*Ha-Boker*], הצופה [*Ha-Tzofeh*], דבר [*Davar*], May 10, 1944.

172. Complete text of the October 3, 1943, ordinance appears in Molho, *In Memoriam*, 189–90.

173. Sevillias, *Athens-Auschwitz*, 6. Etz Hayyim was the "old" synagogue on Melidoni Street.

174. Molho, *In Memoriam*, 191–92, 206–7.

175. Interview with Leon Koen, in Greif, בכינו בלי דמעות, 327.

176. Sevillias, *Athens-Auschwitz*, 7.

177. Personal communication, Sol Kofinas, September 12, 2004.

178. Molho, *In Memoriam*, 208–11.

179. Information of May 1944, Athens, in Constantopoulou and Veremis, *Documents on the History of the Greek Jews*, 278.

180. Molho, *In Memoriam*, 211–12.

181. Mazower, *Inside Hitler's Greece*, 252.

182. *New Judaea* 20, nos. 9–10 (June–July 1944): 143.

183. Interview with Yitzhak Koen in Refael, בנתיבי שאול, 247.

184. See Matsas, *The Illusion of Safety*, 98. For an account of the resistance that gives attention to ties between resistance groups and Greek Jews, see

Ioanni N. Margari, *Η Εποποϊα της Εθνικής Αντίστασις 1941–1944* [The Epic of National Resistance, 1941–1944] (Athens, 1997).

185. For an overview, see Esdra D. Mousi, *Η Εβραϊκή Κοινότητα της Λάρισσας πριν και μετά το Ολοκαύτωμα* [The Jewish Community of Larissa before and after the Holocaust] (Larissa, 2000).

186. Molho, *In Memoriam*, 213–18.

187. Testimony of the sisters Stella Mordohai Miyioni and Eftihia Miyioni, and testimony of Naoum Negrin, in Nahman, *Γιάννενα Ταξίδι στο Παρελθών* (Athens, 1996), 134–39.

188. Testimony of the sisters Stella Mordohai Miyioni and Eftihia Miyioni, in Nahman, *Γιάννενα Ταξίδι στο Παρελθών*, 134–38.

189. Rivlin, פנקס הקהילות, 140. This account is widely corroborated in testimonies and memoirs.

190. Molho and Nehama, שואת יהודי יוון, 136. Molho and Nehama argue that the Romaniote communities of Greece were overly confident that their Greekness would save them, and that they boasted that they were "no different from other Greeks." Because of this hubris, according to Molho and Nehama, the Romaniotes "committed suicide," leaving themselves easy prey for the Germans. See, for example, ibid., 121–22.

191. Rivlin, פנקס הקהילות, 140.

192. U.S. National Archives, No. 1746 (R-1618). Also cited in Matsas, *The Illusion of Safety*, 104–05.

193. Leon Matsa, cited in Εβραϊκή Εστία (Athens), April 16, 1953. Also cited in Ampatzopoulou, *Το Ολοκαύτωμα στης Μαρτυρίες των Ελλήνων Εβραίων*, 235–37.

194. For a complete list, see Marcia Haddad Ikonomopoulos, *In Memory of the Jewish Community of Jannina* (New York, 2004). Molho gives the figure as 1,860, in *In Memoriam*, 218–20.

195. Dalven, *The Jews of Jannina* (Philadelphia, 1990), 43–44. On Kabilli, *Ισραηλιτικόν Βήμα*, no. 7, January 1946. Cited also in Ampatzopoulou, *Το Ολοκαύτωμα στης Μαρτυρίες των Ελλήνων Εβραίων*, 231–32.

196. Testimony of Nina Hor Matsa, in Nahman, *Γιάννενα Ταξίδι στο Παρελθών*, 138.

197. Testimony of Empis Svolis, in Nahman, *Γιάννενα Ταξίδι στο Παρελθών*, 133–34.

198. In Sevillias, *Athens-Auschwitz*, 97–98n2.

199. Markos Nahon, *Μπίρκεναου το Στρατόπεδον του Θανάτου*, [Birkenau, the Death Camp], trans. Asser Moissis (Thessaloniki, 1991), 86.

200. Amariglio, *From Thessaloniki to Auschwitz*, 53.

CHAPTER 8. AUSCHWITZ-BIRKENAU

1. Greek Jews were sent to an array of camps, but primarily Auschwitz-Birkenau, Mauthausen, Treblinka, and Dachau. The vast majority of those who died were gassed in Auschwitz-Birkenau. In Auschwitz, a distinct Greek Jewish type emerged in the minds of other Jews from around Europe.

2. Giakovos Hantali, *Από το Λευκό Πύργο στης Πύλες του ʾΑουσβιτς* (Thessaloniki, 1996), 77.

3. In early 1944, the American War Refugee Board set the number of Jews exterminated at Birkenau as 1.765 million. *New Judaea* 21, nos. 5–6 (February–March 1945): 73. By May 1945, the "conservative" estimate was set at 4 million. *New Judaea* 21, no. 8 (May 1945): 115. Subsequent studies, primarily the work of the Polish historian Franciszek Piper, have established the number of those killed at between 1.1 and 1.5 million, a figure now accepted by the Auschwitz-Birkenau memorial. For an overview, see Sybille Steinbacher, *Auschwitz: A History* (New York, 2005), 132–36.

4. Hantali, *Από το Λευκό Πύργο στης Πύλες του ʾΑουσβιτς*, 87–88.

5. Primo Levi, *If This Is a Man*, trans. Stuart Woolf (London, 1996), 85. (Later published as *Survival in Auschwitz*.)

6. For a discussion of ethnicity in Auschwitz, see Panagis Panagiotopoulos, "*Η χρήσις τής εθνικής αναφοράς στο ʾΑουσβιτς*" [The Deployment of Nationality in Auschwitz], in *Ο Ελληνικός Εβραϊσμός* [Greek Judaism] (Athens, 1998), 99–111.

7. Interview with Mano Avraham Ben-Yaakov, in Shmuel Refael, ed., יוון בשואה יהודי: שאול בנתיבי [Road to Hell: Greek Jews in the Holocaust] (Jerusalem, 1988), 317.

8. Sam Profeta, "Thessaloniki-Auschwitz," *To Dentro*, nos. 37–38 (March–April, 1988), excerpted in Fragkiski Ampatzopoulou, ed., *Το Ολοκαύτωμα στης Μαρτυρίες των Ελλήνων Εβραίων* (Thessaloniki, 1993), 145.

9. Testimony of Leon Koen, in יהדות יוון בחורבנה: פרקי זכרונות [Greek Jewry in the Holocaust: Memoirs] (Tel Aviv, 1988), 91–92.

10. KZ (pronounced in German "ka" "tzet") is the acronym for the German term for concentration camp (*konzentrazionslager*). Inmates were referred to by their "katzetnik number." After the Holocaust, Finer (best known as Yeḥiel Dinur), katzetnik number 135633, took as his name "Ka Tzetnik," saying that after his experience in the camps, he no longer had any other identity. As Primo Levi writes, "Who could tell one of our faces from the other? For [the civilians] we are 'Kazett,' a singular neuter word." *If This Is a Man*, 127.

11. Ka Tzetnik 135633, *Atrocity* (New York, 1963), 132, 146.

12. Hantali, *Από το Λευκό Πύργο στης Πύλες του ʾΑουσβιτς*, 94.

13. Interview with Esther Maestro, in Refael, שאול בנתיבי, 292.

14. Είμαι ένας άνθρωπος απλός εγενήτηκα [sic] στήν Αθήνα στήν παλιά καλή εποχή τό 1901. Errikos Sevillias, *Athens-Auschwitz*, trans. Nikos Stavroulakis (Athens, 1983), frontispiece.

15. Ibid., 33, 35.

16. Erika Kounio Amariglio, *From Thessaloniki to Auschwitz and Back: Memories of a Survivor from Thessaloniki* (London, 2000), 92.

17. Hantali, *Από το Λευκό Πύργο στης Πύλες του ʾΑουσβιτς*, 111.

18. Interview with Alberto Levy, in Refael, שאול בנתיבי, 265.

19. Interview with Shaul Chazan, in Gideon Greif, ed., היהודים מאושוויץ בכינו בלי דמעות: עדויותיהם של אנשי הזונדרקומאנדו [We Wept without Tears: Testimonies of the Jewish *Sonderkommando* from Auschwitz] (Tel Aviv, 1999), 298.

20. Interview with Leon Koen in Greif, בכינו בלי דמעות, 329.

21. Interview with Shabtai Hannuka, in Refael, בנתיבי שאול, 201.

22. Interview with Mano Avraham Ben-Yaakov, in Refael, בנתיבי שאול, 319.

23. Interview with Eliki Sardas, in Refael, בנתיבי שאול, 388.

24. Anonymous, to the tune of the Greek hit "Étsi eínai i zoí" [That's the Way Life Goes], Gianni Bella and Antoni Plomariti. In Almpertos Nar, "Κειμένη επί ακτής θαλάσσης" [Writings from the Edge of the Sea] (Thessaloniki, 1997), 220.

25. Elie Wiesel, introduction to Hantali, Από το Λευκό Πύργο στης Πύλες του 'Αουσβιτς, 12.

26. Amariglio, From Thessaloniki to Auschwitz, 57.

27. Jacques Stroumsa, Violinist in Auschwitz: From Salonika to Jerusalem, 1913–1967, ed. Erhard Roy Wiehn, trans. James Stewart Brice (Konstanz, Ger., 1996). For the original version, see the following note.

28. Iakovos Stroumsa, Διάλεξα τήν Ζωῶ ... Από τή Θεσσαλονίκη στό 'Αουσβιτς [I Chose Life: From Thessaloniki to Auschwitz] (Thessaloniki, 1997), 48–50. Stroumsa was inmate number 121097.

29. Dario Akounis, in Refael, בנתיבי שאול, 40.

30. Primo Levi, The Truce: A Survivor's Journey Home from Auschwitz, trans. Stuart Woolf (London, 1996), 213; see also 207–29.

31. Levi, If This Is a Man, 151.

32. Cited in David Recanati, ed., זכרון שלוניקי: גדולתה וחורבנה של ירושלים דבלקן [Memoir of Salonika: The Glory and Destruction of Jerusalem of the Balkans] (Tel Aviv, 1972), 2:591.

33. Yaakov Razon, in Refael, בנתיבי שאול, 455–56. The similarity between Razon's and Victor Klemperer's observations on the German love of boxing is noteworthy: "[Linguistically] Nazism . . . was influenced by boxing more than all [other kinds of sport] put together. . . . [Its] incarnation of heroism [is] the glassy stare which expresses a hard and thrusting determination coupled with the will to succeed." Victor Klemperer, The Language of the Third Reich. LTI: Lingua Tertii Imperii. A Philologist's Notebook, trans. Martin Brady (London, 2000), 4. A movie, Triumph of the Spirit, was made about a Greek boxer, Shlomo Aroch, in 1990; soon thereafter Jacques Razon filed suit against Aroch; he was the famed boxer of Auschwitz, and not Aroch. Aroch had stolen his story. See Jerusalem Post, May 26, 1990, 12.

34. Interview with Leon Koen, in Greif, בכינו בלי דמעות, 329.

35. Leon Koen, in יהדות יוון בחורבנה, 83.

36. Testimony of Palompas Allalouf, in Erika Kounio Amariglio and Almpertos Nar, Προφορικές Μαρτυρίες Εβραίων της Θεσσαλονίκης για το Ολοκαύτωμα (Thessaloniki, 1998), 40.

37. Leon Parachia, Μαζάλ, αναμνήσεις από τα στρατόπεδα τού θανάτου (1943–1954) [Mazal: Memories from the Death Camps] (Thessaloniki, 1990). Excerpted in Ampatzopoulou, Το Ολοκαύτωμα στης Μαρτυρίες των Ελλήνων Εβραίων, 163. "The Greeks who worked in the crematoriums knew from talking to the others who worked there that at the designated hour they would [also] face the crematoriums."

38. Berry Nahmia, Κραυγή γιά το Αύριο [A Cry for Tomorrow] (Athens, 1989), 157–58.

39. Testimony of Sol Kazes, in Kounio Amariglio and Nar, Προφορικές Μαρτυρίες Εβραίων της Θεσσαλονίκης για το Ολοκαύτωμα, 74–75.

40. From whom they'd just been separated in the selection made on arrival.

41. Testimony of Eftyhia Svoli, in Eftyhia Nahman, Γιάννενα Ταξίδι στο Παρελθών (Athens, 1996), 143.

42. Interview with Eliki Sardas, in Refael, בנתיבי שאול, 388.

43. Interview with Yaakov Gabai, in Greif, בכינו בלי דמעות, 212–20.

44. Ibid.

45. Sevillias, *Athens-Auschwitz*, 41.

46. "Their life is short, but their number is endless; they, the *muselmanner*, the drowned, form the backbone of the camp, an anonymous mass, continually renewed and always identical, of non-men who march and labor in silence, the divine spark dead within them. . . . They crowd my memory with their faceless presences, and if I could enclose all the evil of our time in one image, I would choose this image which is familiar to me: an emaciated man, with head dropped and shoulders curved, on whose face and in whose eyes not a trace of a thought is to be seen." Levi, *If This Is a Man*, 96.

47. Interview with Yaakov Gabai, in Greif, בכינו בלי דמעות, 212–20.

48. Interview with Shaul Chazan, in Greif, בכינו בלי דמעות, 303.

49. Ibid.

50. Interview with Emmanuel Hannuka, in Refael, בנתיבי שאול, 195.

51. Interview with Leon Koen, in Greif, בכינו בלי דמעות, 329.

52. Amariglio, *From Thessaloniki to Auschwitz*, 90–91.

53. Sevillias, *Athens-Auschwitz*, 50–52.

54. Ibid., 104n3.

55. Leon Koen, "The Rebellion of the Sonderkommando," in יהדות יוון בחורבנה, 101.

56. Summary account of the rebellion are drawn from Grief, בכינו בלי דמעות, 74–78. The 212 figure is disputed, and is certainly high. Reports as to the number of workers who survived the revolt vary wildly.

57. Koen, "The Rebellion," 106. See also Sevillias, *Athens-Auschwitz*, 50–54; 104–5. David Persiades, Sevillias's brother-in-law, participated in the revolt.

58. Sevillias, *Athens-Auschwitz*, 40–42, 103.

59. Koen, "The Rebellion," 106.

60. Testimony of Albertos Menasche, in Ampatzopoulou, Το Ολοκαύτωμα στης Μαρτυρίες των Ελλήνων Εβραίων, 71. A doctor before and after the war, Menasche was assigned number 124454 in Auschwitz. His testimony was recorded in 1945.

61. Testimony of Esthir Iakov, in Nahman, Γιάννενα Ταξίδι στο Παρελθών, 147. On the sterilization of men, see also the "anonymous" testimony of Leon Kapon at the Eichmann trial in 1961.

62. Testimony of Esthir Iakov.

63. Markos Nahon, Μπίρκεναου το Στρατόπεδον του Θανάτου [Birkenau, the Death Camp] (Thessaloniki, 1991), 74. Nahon was sent on a transport on May 10, 1943, from Salonika to Auschwitz, where he was assigned inmate number 122274.

64. Interview with Leon Koen, in Greif, בכינו בלי דמעות, 335–36; 328.

65. Nahon, *Μπίρκεναου το Στρατόπεδον του Θανάτου*, 75–76.

66. Hantali, *Από το Λευκό Πύργο στης Πύλες του 'Αουσβιτς*, 86.

67. Amariglio, *From Thessaloniki to Auschwitz*, 88.

68. American Joint Distribution Committee representative Israel G. Jacobson to presidents' meeting, National Council of Jewish Women, February 27, 1946.

69. Amariglio, *From Thessaloniki to Auschwitz*, 88.

70. Testimony of Elviras Kolanto, in Kounio Amariglio and Nar, *Προφορικές Μαρτυρίες Εβραίων της Θεσσαλονίκης για το Ολοκαύτωμα*, 90–91

71. Testimony of Oro Alfantari, in Kounio Amariglio and Nar, *Προφορικές Μαρτυρίες Εβραίων της Θεσσαλονίκης για το Ολοκαύτωμα*, 48–51.

72. Amariglio, *From Thessaloniki to Auschwitz*, 100.

73. Testimony of Naoum Negrin, in Nahman, *Γιάννενα Ταξίδι στο Παρελθών*, 140–41.

74. Interview with Esther Rafael, in Refael, בנתיבי שאול, 466.

75. Interview with Yitzhak Koen, in Refael, בנתיבי שאול, 247. See also the testimony of Albertos Menasche in Ampatzopoulou, *Το Ολοκαύτωμα στης Μαρτυρίες των Ελλήνων Εβραίων*, 76–79. There is confusion on the chronology; the Athens transport more likely arrived in May.

76. Greif, בכינו בלי דמעות, 220. Further transports came from Rhodes, which at the time was not part of Greece.

77. In Sevillias, *Athens-Auschwitz*, 104n4.

78. For a meticulously reconstructed chronology of the daily happenings in Auschwitz, see Danuta Czech, *Auschwitz Chronicle, 1939–1945* (New York, 1990).

79. *'Η Καθημερινή*, February 1, 1945.

80. Leon Parachia, *Μαζάλ, αναμνήσεις από τα στρατόπεδα τού θανάτου (1943–1954)*, excerpted in Ampatzopoulou, *Το Ολοκαύτωμα στης Μαρτυρίες των Ελλήνων Εβραίων*, 163. "Ενιωθαν 'Ελληνες καί πεθάναν 'Ελληνες."

CHAPTER 9. TRYING TO FIND HOME: JEWS IN POSTWAR GREECE

1. *New Judaea* 21, no. 3 (December 1944): 38.

2. PRO/FO 371/51171 98605, reg. no. WR 496/274/48, Foreign Office Refugee Department to the British Legation, March 14, 1945, Berne, Switzerland.

3. On Salonika immediately after the war, see Bea Lewkowicz, "'After the War We Were All Together': Jewish Memories of Postwar Thessaloniki," in *After the War Was Over: Reconstruction of the Family, Nation, and State in Greece, 1943–1960* (Princeton, NJ, 2000), 247–72.

4. When he arrived in Jannina, he found forty-two Jewish male survivors, all between the ages of eighteen and twenty-eight, and five females. Forty-eight hours later, having found only one surviving cousin, Levy made his way back to Patras. The Odyssey of Louis Levy, May 31–June 3, 1945, oral history and exhibit, summer 2004, Kehila Kedosha Janina Synagogue and Museum, New York.

5. See PRO/FO 371/51171, reg. no. WR 452/274/48ff.

6. PRO/FO 371/51171 98605, reg. no. WR 496/274/48, enclosure, Letter of I. Nefussy to British Embassy, January 20, 1945, Athens.

7. Ibid.
8. Menachem Koen, Allalouf, and Soulema of Salonika; Kanti and Tarampolou of Didimoticho; and Kalbo from Nea Orestiada.
9. Isaak Matarasso, "Κι όμως όλοι τους δεν πέθαναν" [And Yet They Didn't All Die] (Athens, 1948), excerpted in Fragkiski Ampatzopoulou, ed., *To Ολοκαύτωμα στης Μαρτυρίες των Ελλήνων Εβραίων* (Thessaloniki, 1993), 118, 121–27.
10. Interview with Ovadia Baruch, in Shmuel Refael, ed., שאול: יהודי יוון בשואה בנתיבי [Road to Hell: Greek Jews in the Holocaust] (Jerusalem, 1988), 104.
11. Interview with Aliza Baruch, in Refael, שאול בנתיבי, 112–13.
12. AJDC representative Israel G. Jacobson to presidents' meeting, National Council of Jewish Women, February 27, 1946.
13. Interview with Freida Kobo, in Refael, שאול בנתיבי, 436–37.
14. *Εβραϊκή Εστία*, April 16, 1953, Athens, cited in Ampatzopoulou, *To Ολοκαύτωμα στης Μαρτυρίες των Ελλήνων Εβραίων*, 235–37.
15. Lewkowicz, "After the War We Were All Together," 264.
16. See, for example, "Δύο μέρες στο δικαστήριο" [Two Days in Court], *Ισραηλιτικόν Βήμα*, July 5, 1946. While quick to ferret out Communists, the postwar Greek government was lax in the pursuit of Nazi collaborators. Few Greek Christian war criminals were ever brought to trial, despite the fact that in many cases their names and whereabout were well-known. In a trial held during May 1945, high-ranking members of the puppet government, among them Tsolakoglou, Gotsamanis, and Tsironikos, were sentenced to death (Gotsamanis and Tsironikos in absentia), but hundreds of known collaborators went free. A handful of civilian collaborators were also given the death sentence, but since the verdicts were rendered in absentia, they were able to flee the country. (*To Βήμα*, June 1, 1945; see also Michael Matsas, *The Illusion of Safety: The Story of the Greek Jews during the Second World War* [New York, 1997], 406.) Jewish collaborators were tracked down by other Jews, not by the Greek government. For example, Vital Hasson, who had served as leader of the Nazi-organized Jewish police force in Salonika, was given help by the Italians in his escape to Albania. He later escaped to Bari, Italy. When he was recognized there by a fellow Salonikan and reported to the Allied authorities, he was detained briefly but then released. After later being arrested by the British in Egypt, he was sent to Athens and again released. Only after returning to Salonika, where he was recognized, beaten, and turned over to the police, did he end up in prison. On Hasson, see the earliest edition of Michael Molho, *In Memoriam: Hommage aux Victimes Juives des Nazis en Grece* (Thessaloniki, 1948), 1:91–92.
17. The Recanati brothers, Ino and Epo, along with Daniel Koen, Jacques Alballa, Leon Sion, and Edgar Kounio were tried in Greece in June 1947. Hasson was given a death sentence and killed by a firing squad; the others served prison terms ranging from eight years to life.
18. Rozina Asser-Pardo, *548 Ήμέρες με Άλλο Όνομα. Θεσσαλονίκη 1943. Μνήμες πολέμου* [548 Days with a Different Name: Salonika 1943 War Memories] (Athens, 1999), 12–13.
19. Confidential report of G. A. Christodoulou, Consul for Palestine and Trans-Jordan, to the Greek Ministry of Foreign Affairs, June 27, 1945, Jerusalem,

in Photini Constantopoulou and Thanos Veremis, eds., *Documents on the History of the Greek Jews: Records from the Historical Archives of the Ministry of Foreign Affairs* (Athens, 1998), 325–27.

20. Personal communication with Elli Koen, October 1999.

21. *Ισραηλιτικόν Βήμα*, no. 1, November 23, 1945. Cited also in Ampatzopoulou, *Το Ολοκαύτωμα στης Μαρτυρίες των Ελλήνων Εβραίων*, 229–30.

22. Testimony of Sol Kazes, in Erika Kounio Amariglio and Almpertos Nar, *Προφορικές Μαρτυρίες Εβραίων της Θεσσαλονίκης για το Ολοκαύτωμα* (Thessaloniki, 1998), 69.

23. *Ισραηλιτικόν Βήμα*, no. 1, November 23, 1945. Cited also in Ampatzopoulou, *Το Ολοκαύτωμα στης Μαρτυρίες των Ελλήνων Εβραίων*, 230.

24. AJDC report on Greece (Marvin Goldfine), European Executive Council Research Department, February 3, 1947.

25. *New Judaea* 21, no. 3 (December 1944): 38.

26. AJDC representative Israel G. Jacobson to presidents' meeting, National Council of Jewish Women, February 27, 1946.

27. AJDC report on Greece (Marvin Goldfine), European Executive Council Research Department, February 3, 1947; see also *New York Herald Tribune*, December 18, 1955.

28. See, for example, "Καινούργιες Οικογένειες" [New Families], *Ισραηλιτικόν Βήμα*, April 15, 1946.

29. Koula Cohen Kofinas, oral history recorded by Kehila Kedosha Janina, New York.

30. AJDC report on Greece (Marvin Goldfine), European Executive Council Research Department, February 3, 1947.

31. AJDC representative Israel G. Jacobson to presidents' meeting, National Council of Jewish Women, February 27, 1946.

32. Cited in *Philhellene* 1, no. 1 (January 1, 1942): 1.

33. On out-migration from the islands during the early stages of the occupation, see Violetta Hionidou, *Famine and Death in Occupied Greece, 1941–1944* (Cambridge, 2006), 148ff. By some reports the number of destroyed villages was much higher, in the thousands. See, for example, A. A. Pallis, *Reconstruction in Greece since the Liberation, 1945–1946* (London, 1947).

34. Pallis, *Reconstruction in Greece*.

35. AJDC representative Israel G. Jacobson to presidents' meeting, National Council of Jewish Women, February 27, 1946.

36. PRO/FO 371/52507 110588, Jewish Refugees in Greece, February 4, 1946.

37. PRO/FO 286/1179 99309, ref. 100/3/7/12, Minister Mercouris, Kingdom of Greece, Sub Ministry of Public Security, Direction of Aliens Service, Section B, February 20, 1946, Athens.

38. PRO/FO 286/1178 99309, 50/37/46, Top Secret, Henley to Reilly, Jewish Affairs, March 18, 1946. Jewish communities in Greece on occasion sought the permission of the Greek government to take in Jewish refugees, and had to promise to cover all costs associated with them and guarantee that they would be law abiding. See, for example, PRO/FO 286/1178 99309, H. Schachnai, Jewish National Council for Palestine to Greek Minister for Public Security, February 11,

1946, Athens; PRO/FO 286/1178 99309, H. Schachnai to Minstry of Public Order, February 6, 2006, Athens. The instances in which it was allowed were exceedingly rare.

39. AJDC representative Israel G. Jacobson to presidents' meeting, National Council of Jewish Women, February 27, 1946. Also see Lewkowicz, "After the War We Were All Together," 259.

40. PRO/FO 286/1178 99309, no. 463/5/3/46–50/78/46G, British Consulate, Corfu to British Embassy, May 13, 1946, Athens; PRO/FO 286/1178 99309, no. 469/130/46, British Consulate, Salonika, to British Embassy, May 28, 1946, Athens.

41. AJDC representative Israel G. Jacobson to presidents' meeting, National Council of Jewish Women, February 27, 1946.

42. *New Judaea* 20, no. 12 (September 1944): 90.

43. *New Judaea* 21, no. 3 (December 1944): 38.

44. See, for example, Ministerial Decision of M. Pesmazoğlu, Minister of Finance, May 31, 1945, Athens, in Constantopoulou and Veremis, *Documents on the History of the Greek Jews*, 319–25.

45. *Government Gazette* (Athens) 1, no. 11 (November 10, 1944), in Constantopoulou and Veremis, *Documents on the History of the Greek Jews*, 307–9.

46. J. Romanos, Counsellor, Embassy in London to Greek Ministry of Foreign Affairs, September 3, 1945, London, in Constantopoulou and Veremis, and attachments, *Documents on the History of the Greek Jews*, 329–34.

47. "The Greek State which, under our existing law, has the right of succession to all property left without heirs, now relinquishes this right with respect to . . . Jewish property, and transfers this right to a specially-planned Foundation whose aim shall be to relieve and rehabilitate our Jewish countrymen who have survived." Giorgios Mavros, undersecretary of state to the prime minister and minister of justice, Greece, to Israel Jacobson, AJDC, January 17, 1946. Also appears as document 123, in Constantopoulou and Veremis, *Documents on the History of the Greek Jews*, 343–46. (A. Kyrou, director, Ministry of Foreign Affairs, to the Greek Consulate, Boston, January 15, 1946, Athens.)

48. *Οργανισμός Περιθάλψεος και Αποκαταστάσεος Ισραηλίτων Ελλάδος, ΟΠΑΙΕ* (OPAIE), the Agency for the Rehabilitation and Care of Greek Jews, established for the restitution of Greek Jewish property, remains to this day the central Greek Jewish political institution.

49. AJDC representative Israel G. Jacobson to presidents' meeting, National Council of Jewish Women, February 27, 1946. For similar accounts, see also Rae Dalven, *The Jews of Ioannina* (Philadelphia, 1990), 49; Lewkowicz, "After the War We Were All Together," 260.

50. Interview with David Mordoch, in Refael, בנתיבי שאול, 302.

51. *Ισραηλιτικόν Βήμα*, no. 1, November 23, 1945. Cited also in Ampatzopoulou, *Το Ολοκαύτωμα στης Μαρτυρίες των Ελλήνων Εβραίων*, 230–31.

52. Asser-Pardo, *548 Ήμέρες με Άλλο Όνομα*, 71.

53. *Ισραηλιτικόν Βήμα*, no. 7, January 1946. Cited also in Ampatzopoulou, *Το Ολοκαύτωμα στης Μαρτυρίες των Ελλήνων Εβραίων*, 231–32.

54. *Εβραϊκή Εστία* (Athens), July 4, 1947. Cited also in Ampatzopoulou, *Το Ολοκαύτωμα στης Μαρτυρίες των Ελλήνων Εβραίων*, 235.

55. *Ισραηλιτικόν Βήμα*, no. 7, January 1946. Cited also in Ampatzopoulou, *Το Ολοκαύτωμα στης Μαρτυρίες των Ελλήνων Εβραίων*, 233–34.

56. AJDC representative Israel G. Jacobson to presidents' meeting, National Council of Jewish Women, February 27, 1946.

57. AJDC Athens (Harold Goldfarb) to AJDC New York (Robert Pilpel), March 20, 1950.

58. *Ισραηλιτικόν Βήμα*, no. 7, January 1946. Cited also in Ampatzopoulou, *Το Ολοκαύτωμα στης Μαρτυρίες των Ελλήνων Εβραίων*, 233–34.

59. AJDC country director's report on Greece, letter no. 670, September 5, 1950.

60. Hionidou, *Famine and Death in Occupied Greece*, 109–47.

61. ΚΙΣ (KIS, Central Board of Jewish Communities of Greece), report no. 13, "*Δράσις κατά μήνα φεβρουάριον 1948*" [Activity during February 1948], March 12, 1948, Athens.

62. Minutes, meeting of the Administrative Committee of the AJDC, May 13, 1947.

63. PRO/FO HS 5/351 102373, M. Ward, Major, to HQ Force 133, April 26, 1945.

64. AJDC representative Israel G. Jacobson to presidents' meeting, National Council of Jewish Women, February 27, 1946.

65. Beckelman to Paris headquarters, AJDC, letter no. 5418, July 23, 1947.

66. AJDC representative Israel G. Jacobson to presidents' meeting, National Council of Jewish Women, February 27, 1946.

67. PRO/FO 371/52507 110588, Jewish Refugees in Greece, February 4, 1946.

68. PRO/FO 371/45405 108100, ΚΙΣ, November 23, 1945, Athens.

69. AJDC country director's report on Greece, letter no. 670, September 5, 1950.

70. A. Menascé (president of the Jewish community of Salonika) to Harold Goldfarb (director of the AJDC), September 22, 1947.

71. Sephardic Brotherhood of America, Inc., to the AJDC, July 11, 1947, New York.

72. Ibid.

73. *JTA Bulletin*, March 9, 1950.

74. AJDC report on Greece (Marvin Goldfine), European Executive Council Research Department, February 3, 1947.

75. AJDC representative Israel G. Jacobson to presidents' meeting, National Council of Jewish Women, February 27, 1946.

76. AJDC country director's report on Greece, letter no. 670, September 5, 1950.

77. AJDC Paris to AJDC New York, letter no. 2764, July 14, 1947.

78. AJDC report on Greece (Marvin Goldfine), European Executive Council Research Department, February 3, 1947; UNRRA HQ, Greece Mission, Welfare Division, Harry Greenstein, director of welfare, Balkan mission, to Reuben Resnik, Intergovernmental Committee of Refugees, February 23, 1945.

79. AJDC country director's report on Greece, letter no. 670, September 5, 1950.

80. PRO/FO 371/52507 110588, Jewish Refugees in Greece, February 4, 1946.

81. PRO/FO 286/1178 99309, 100/3/7/36, Athens Ministry of Public Order to the British Police Mission, July 30, 1946; PRO/FO 286/1178 99309, 50/141/46, UNRRA (Stokes) to British Embassy (Reilly), August 6, 1946, Athens.

82. PRO/FO 286/1178 99309, H. Schachnai (Jewish National Council for Palestine) to Minister of Public Security (Mercouris), February 11, 1946.

83. PRO/FO 286/1178 99309, 100/3/7/36, Athens Ministry of Public Order to the British Police Mission, July 30, 1946. For a full list of the names of this sole group of legally admitted international refugees, provided by UNRRA, see PRO/FO 286/1178 99309, 50/22/46, British High Commissioner for Palestine (Cunningham) to British Ambassy, March 5, 1946, Athens. PRO/FO 286/1178 99309, Th. Sophoulis, Greek Prime Minister (as Minister for Public Security), order 100/3/7/12, March 15, 1946, Athens; PRO/FO 286/1178 99309, 50/37/46, British Embassy, Reilly to Henley re: Jewish Affairs, Athens.

84. Mercouris, the Greek minister of public security, wrote of "the danger of [non-Greek Jewish refugees] being registered as members of the Jewish community of Salonika." PRO/FO 286/1178 99309, ref. 100/3/7/12, February 20, 1946, Athens.

85. PRO/FO 286/1178 99309, 100/3/7/36, Athens Ministry of Public Order to the British Police Mission, July 30, 1946.

86. PRO/FO 286/1178 99309, no. 463/5/3/46–50/78/46G, British Consulate, Corfu to British Embassy, May 3, 1946, Athens.

87. PRO/FO 286/1178 99309, 50/82/46, British Consulate, Patras (Harcourt) to British Embassy, (Reilly), May 22, 1946, Athens.

88. PRO/FO 286/1178 99309, British Consul, Salonika (Peck) to British Embassy (Reilly), Salonika, January 9, 1946, Athens.

89. PRO/FO 286/1178 99309, 50/5/46, Σιωνιστική Ομοσπονδία της Ελλάδος [Zionist Federation of Greece], January 10, 1946, Athens. The declaration wasn't deemed important enough for the British Embassy in Athens to pass it on to the home office; scrawled in the margin is the ironic note, "As this merely says that the Zionists in Greece are in favour of Zionism I doubt if it is worth sending home."

90. ΚΙΣ, report no. 47, "Δράσις κατά μήνα φεβρουάριον 1948," December 1948, Athens.

91. ΚΙΣ, report no. 49, "Δράσις κατά μήνα φεβρουάριον 1948" [Activity in December 1948], January 5, 1949, Athens.

92. Interview with Yitzhak Bileli, in Refael, בנתיבי שאול, 72.

93. Matsas, The Illusion of Safety, 416–19.

94. For transcripts of some of the proceedings, see Le Procès d'Espionnage en Grèce (15 Fevrier–1er Mars 1952) (Athens, 1952).

95. Terrorism in Greece: Αυτό είναι το ΚΚΕ-ΕΑΜ-ΕΛΑΣ στην Ελλάδα [This Is the KKE-EAM-ELAS in Greece] (Athens, 1945), frontispiece, 35, 38.

96. See, for example, Christophoros Chrestides, La Tradition Democratique de la Grèce: L'aspect réel de la situation actuelle (Athens, 1948), 6

97. By Greek government estimates 21,800 former Communists had been "processed" by Makronissos by 1949. For an account of Civil War political pris-

oners, see Polymeris Voglis's superb *Becoming a Subject: Political Prisoners during the Greek Civil War* (New York, 2002).

98. C. P. Rodocanachi, *A Great Work of Civic Readaptation in Greece* (Athens, 1949), 7, 12, 17. Of the 21,800 mentioned in the previous note, 800 had been deemed "incorrigible"—meaning, that they would not under any circumstances renounce their Communism and were not susceptible to "readaptation."

99. Interview with Elimelech Baruch, in Refael, בנתיבי שאול, 91.

100. Interview with Shaul Ben-Maor, in Refael, בנתיבי שאול, 85.

101. PRO/FO 286/1178 99309, no. 463/5/3/46–50/78/46G, British Consulate, Corfu, to British Embassy, May 13, 1946, Athens.

102. In summer 1949, Jewish Communists sought an official Cominform statement against Zionism—a move strongly supported by the Greek Jewish Left. *Jewish Newsnotes* 1, no. 25 (July 1, 1949), 2.

103. G. A. Christodoulou, Consul for Palestine and Trans-Jordan, to the Greek Ministry of Foreign Affairs, September 23, 1945, Jerusalem, in Constantopoulou and Veremis, *Documents on the History of the Greek Jews*, 340–41.

104. Interview with Gedalia Levy, in Refael, בנתיבי שאול, 282.

105. Interview with Isidor Alalouf in Refael, בנתיבי שאול, 35.

106. Interview with Yaakov Jabari, Refael, בנתיבי שאול, 137.

107. Interview with Alberto Gatenio, Refael, בנתיבי שאול, 143–44.

108. Political Office of the Prime Minister, Minutes of Meeting between Prime Minister and Shertock (Attachment), April 15, 1945, in Constantopoulou and Veremis, *Documents on the History of the Greek Jews*, 318–19.

109. D. Papas, Ambassador of Greece in Cairo, to the Greek Ministry of Foreign Affairs, September 18, 1944, Cairo, in Constantopoulou and Veremis, *Documents on the History of the Greek Jews*, 299–300. A second precondition for accepting the creation of a Jewish state was "that none of the rights of the non-Jewish populations of Palestine . . . be harmed."

110. G. A. Christodoulou, Consul for Palestine and Trans-Jordan, to the Greek Ministry of Foreign Affairs, November 21, 1944, Jerusalem, in Constantopoulou and Veremis, *Documents on the History of the Greek Jews*, 309–10.

111. G. A. Christodoulou, Consul for Palestine and Trans-Jordan, to the Greek Ministry of Foreign Affairs, September 12, 1945, Jerusalem, in Constantopoulou and Veremis, *Documents on the History of the Greek Jews*, 339.

112. AJDC representative Israel G. Jacobson to presidents' meeting, National Council of Jewish Women, February 27, 1946.

113. Confidential Report of G. A. Christodoulou, Consul for Palestine and Trans-Jordan, to the Greek Ministry of Foreign Affairs, June 27, 1945, Jerusalem, in Constantopoulou and Veremis, *Documents on the History of the Greek Jews*, 325–27.

114. ΚΙΣ/ΟΠΑΙΕ/JDC meeting on Jewish property, ΚΙΣ no. 467, August 13, 1952; 44/67.

115. The "Moissis Plan" of 1949, designed by Asher Moissis, then president of the Central Board of Jewish Communities of Greece (ΚΙΣ).

116. Jerome Jacobson, general counsel, AJDC Paris, to Harold Goldfarb, AJDC Athens, letter no. 628, April 21, 1950, Paris.

117. *Jewish Newsnotes* 2, no. 6 (March 3, 1950): 3; *Jewish Newsnotes* 2, no. 8 (March 17, 1950): 2.

118. *Jewish Newsnotes* 2, no. 46 (December 8, 1950): 1.

119. The situation was exacerbated by ongoing negotiations between Turkey, Greece, and the Arab states to establish a Mediterranean defensive pact, of which Greek premier Sophocles Venizélos was a staunch supporter. *Jewish Newsnotes* 2, no. 50 (January 5, 1951): 1.

CHAPTER 10. HELLENIZED AT LAST: GREEK JEWS IN PALESTINE/ISRAEL

1. *Greece, Basic Statistics: 1949* (London, 1949), 16–17.

2. *United Palestine Appeal 1941 Yearbook* (New York, 1941), 22, 24, 71.

3. Israel Office of Information, RP 17, April 1951 ("Three Years of Israel's Statehood: The Story of Immigration"), RP 17/a, April 1952 ("Four Years of Israel's Statehood: The Story of Immigration"), and subsequent annual reports, in vol. 1, *Israel Office of Information Miscellaneous Publications*.

4. Tom Segev, *1949: The First Israelis* (New York, 1986), 95.

5. G. A. Christodoulou, Consul for Palestine and Trans-Jordan, to the Greek Ministry of Foreign Affairs, September 6, 1945, Jerusalem, in Photini Constantopoulou and Thanos Veremis, *Documents on the History of the Greek Jews*, 334–38.

6. Ibid.

7. דבר, August 9, 1945. See also הארץ, "The People Who Return to Greece from Palestine," cited in Constantopoulou and Veremis, *Documents on the History of the Greek Jews*, 336–37.

8. *Palestine and Transjordan* 2, no. 65 (September 11, 1937): 1.

9. *Palestine and Transjordan* 2, no. 66 (September 18, 1937): 6; *New Judaea* 22, no. 9 (June 1946): 163.

10. *Palestine and Transjordan* 2, no. 73 (February 5, 1938): 3; *Palestine and Transjordan* 2, no. 74 (May 7, 1938): 3. See also public security section of the "Annual Report of the Government on the Administration of Palestine and Transjordan, 1937," *Palestine and Transjordan*, 3, no. 80 (June 18, 1938): 4–5.

11. *Palestine and Middle East* 19, no. 2 (February 1947): 39.

12. *Palestine and Middle East* 19, nos. 3–4 (March–April 1947): 51.

13. *New Judaea* 22, no. 9 (June 1946): 163.

14. *Jewish Newsnotes* 3, no. 19 (May 18, 1951): 2.

15. One example of making their mark is the Ginio family, which was known in Jerusalem for its winery. See Alisa Meyuhas Ginio, "The Ginios of Salonika and Wine Production in Jerusalem," in *The Mediterranean and the Jews: Society, Culture, and Economy in Early Modern Times II*, ed. Elliott Horowitz and Moises Orfali (Ramat Gan, Israel, 2002), 157–74.

16. *Palestine and Transjordan* 2, no. 52 (June 12, 1937): 1–2.

17. *Palestine and Transjordan* 1, no. 38 (February 20, 1937): 2–3.

18. Arye Dissenchik, מעריב [*Ma'ariv*], January 7, 1949. Cited in Segev, *1949: The First Israelis*, 119. For a general description of the physical circumstances of immigration, see ibid., 117–54.

19. *United Israel Appeal 1952 Yearbook* (Washington, DC, 1952), 61.

20. *Palestine and Middle East* 17, no. 10 (October 1946): 223–27, 236; *New Judaea* 23, nos. 3–4 (December 1946–January 1947): 74–75.

21. "Tel Aviv: New Phase," *Palestine and Middle East* 19, no 1 (January 1947): 5–7.

22. "The Jewish Case before UNSCOP," *Palestine and Middle East* 19, nos. 6–7 (June–July 1947): 101ff.

23. Giakovos Hantali, *Από το Λευκό Πύργο στης Πύλες του 'Αουσβιτς* (Thessaloniki, 1996), 94.

24. Testimony of Pavlos Simcha, excerpted in Fragkiski Ampatzopoulou, ed., *Το Ολοκαύτωμα στης Μαρτυρίες των Ελλήνων Εβραίων* (Thessaloniki, 1993), 164.

25. Testimony of Esther Pitson, in Ampatzoglou, *Το Ολοκαύτωμα στης Μαρτυρίες των Ελλήνων Εβραίων*, 190.

26. Testimony of Loutsiana Modiano-Soulam, in Ampatzoglou, *Το Ολοκαύτωμα στης Μαρτυρίες των Ελλήνων Εβραίων*, 191.

27. Testimony of Maty Azaria, in Erika Kounio Amariglio and Almpertos Nar, *Προφορικές Μαρτυρίες Εβραίων της Θεσσαλονίκης για το Ολοκαύτωμα* (Thessaloniki, 1998), 22.

28. On the broader relationship between Israeli cinema and ideology, see Ella Shohat, *Israeli Cinema: East/West and the Politics of Representation* (Austin, 1989).

29. The filmmakers are Alfred Steinhardt and Yechiel Niemann.

30. Yitzhak Rafael Molho, ימאים שאלוניקאים בישראל: חזון והגשמה [Salonikan Sailors in Israel: Vision and Realization] (Jerusalem, 1951).

31. *New Judaea* 21, nos. 1–2 (October–November 1944): 23–24.

32. There is a copious literature on the mizraḥi/Ashkenazic divide in Israel and the Orientalization of the mizraḥim. See, among others, Ella Shohat, "The Invention of the Mizraḥim," *Journal of Palestine Studies*, no. 1 (Autumn 1999): 5–20; Ella Shohat, "Sephardim in Israel: Zionism from the Standpoint of Its Jewish Victims," *Social Text* 19/20 (Fall 1988): 1–35; Ella Shohat, "Columbus, Palestine, and Arab-Jews: Toward a Relational Approach to Community Identity," in *Orientalism and History*, ed. Edmund Burke III and Dilip Basu (Ann Arbor, MI, 2001); Moshe Shokeid, "On the Sin We Did Not Commit in the Research of Oriental Jews," *Israel Studies* 6, no. 1 (2001): 15–33; Yehouda Shenhav, "The Phenomenology of Colonialism and the Politics of 'Difference': European Zionist Emissaries and Arab-Jews in Colonial Abadan," *Social Identities* (2002): 521–44; Avner Ben-Amos, "An Impossible Pluralism: European Jews and Oriental Jews in the Israeli History Curriculum," *History of European Ideas* 18 (1994): 41–51; Gabriel Piterberg, "Domestic Orientalism: The Representation of 'Oriental' Jews in Zionist/Israeli Historiography," *British Journal of Middle Eastern Studies* 23 (1996): 125–45.

33. Coproduced by Yona Films and Shapira Films.

34. http://www.geocities.com/Nashville/Opry/3795/main.htm.

35. Eran Riklis, director; written by Amir Ben-David and Moshe Zonder.

36. Motti Regev and Edwin Seroussi, *Popular Music and National Culture in Israel* (Berkeley, CA, 2004), 200.

37. Personal communication, "Levi Levi" (Levitros), November 1, 2006.

38. Significant in this regard is the fact that the only academic context in Israel in which modern Greek is taught is Tel Aviv University's Mediterranean Studies Program (which also plans to offer Ladino, which it describes as "the quintessential Mediterranean Jewish language). See http://www.tau.ac.il/humanities/cmc/study.html.

39. Michal Palti, "Does Pitsa Papadopoulos Ring a Bell?" Ha-Aretz Magazine (English ed.), October 22, 1999.

40. On Mediterranean Music, see Amy Horowitz, "Musika Yam Tikhonit Yisraelit (Israeli Mediterranean Music): Cultural Boundaries and Disputed Territories" (PhD diss., University of Pennsylvania, 1994).

41. Theodorakis frequently characterizes Israeli policy toward Palestinians as "Nazi." See, for example, "The Jewish Problem, according to Theodorakis," Ha-Aretz (English ed.), August 28, 2004; see also "Greek Composer Mikis Theodorakis Is Compared to Hitler after Saying Jews Are 'the Root of All Evil,'" at http://news.bbc.co.uk/go/em/fr/-/1/hi/entertainment/music/3266891.stm.
At Yāsir 'Arafāt's request, Theodorakis composed the Palestinian national anthem, basing it on a song sung by Greek partisans during the resistance to Nazi occupation.

42. Regev and Seroussi, Popular Music and National Culture in Israel, 193. Jason M. Kemper, "Defining a Musical Culture: The Case of Musica Mizrahit," in University of Chicago online journals, humanities.uchicago.edu/jsjournal/kemper.html. See also Jeff Halper, Edwin Seroussi, and Pamela Squires-Kidron, "Musika Mizrahit: Ethnicity and Class Culture in Israel," Popular Music 8, no. 2 (1989): 131–41.

43. See Eliezer Moshe Finegold, Musika Mizrahit: From the Margins to the Mainstream (Cambridge, MA, 1996).

44. Cited in Palti, "Does Pitsa Papadopoulos Ring a Bell?" A fully Greek radio station broadcasting from Tel Aviv is Radio Yasoo, which broadcasts on the Internet at http://www.radio-yasoo.co.il/index.html. Founded by Eliyahu "Yasoo" Zilbermann, the station was inspired by Zilbermann's love of Nikos Iakovides's Lesbos music site, and is particularly dedicated to traditional Greek musical genres.

45. Shmuel Sasson, ed., לא נשכח!, book 6 (1991), 18.

46. Pnina Zakar, ed., לא נשכח!, book 5 (1990), 25.

47. Interview with Mano Avraham Ben-Yaakov (October 16, 1985), in Shmuel Refael, ed., בנתיבי יוון בשואה: יהודי שאול [Road to Hell: Greek Jews in the Holocaust] (Jerusalem, 1988), 321.

48. State Attorney Bar-Or, on report 832, Eichmann trial session 47, in The Trial of Adolf Eichmann (Jerusalem, 1992), 2:847.

49. Trial session 47, in The Trial of Adolf Eichmann, 2:847–48.

50. The spelling of Itzchak Nechama is per the English translation of the trial transcripts.

51. Hanna Yablonka, The State of Israel vs. Adolf Eichman, trans. Ora Cummings with David Herman (New York, 2004), 88–89.

52. Ibid., 186.

53. Eichmann Trial, Prime Minister's Office, Information Center, 440/220, May 18, 1961. Cited in Yablonka, *The State of Israel*, 186.

54. הארץ, April 11, 1961.

55. No witness was provided for Bulgaria, nor was North African Jewry included in the trial in any way. דבר, July 7, 1961. Cited in Yablonka, *The State of Israel*, 188.

56. הצופה, May 30, 1961. Cited in Yablonka, *The State of Israel*, 189.

57. Personal communication, Hyman Genee, July 1999.

CHAPTER 11. CONCLUSION: GREEK JEWISH HISTORY—GREEK OR JEWISH?

1. Established by law as Remembrance Day of Greek Jewish Heroes and Martyrs of the Holocaust. *Εφημερίς της Κυβερνήσεως της Ελληνικής Δημοκρατίας* [Journal of the Greek Democratic Government] 12 (January 27, 2004).

2. Foreign Minister Georgios A. Papandreou, January 27, 2004, speech at Athens Music Mansion, Greek Ministry of Foreign Affairs press release.

3. Katerina Boura, Greek consul-general in New York, Holocaust Remembrance Day, Center for Jewish History, January 27, 2006, New York.

4. *Το Βήμα* [*To Víma*], April 14, 2002. For one angry Israeli response (among many), see Sharon Sadeh, "They Compare Sharon to Hitler," *Ha-Aretz* (English ed.), January 21, 2003. The episode was closely followed by human rights watchdog groups.

5. Personal communication, Marcia Haddad Ikonomopoulos, September 12, 2004.

6. Koula Cohen Kofinas, oral history recorded by the Kehila Kedosha Janina, New York.

7. Heinz D. S. Kounio to the family of Vasilis Rakopoulos, August 9, 2000, Athens. Courtesy of the Kehila Kedosha Janina, New York.

8. Personal communication, Sol Kofinas, September 12, 2004.

9. Ilias Hadjis, oral history recorded by the Kehila Kedosha Janina, New York.

10. From the exhibit Seven Saviours. Courtesy of the Kehila Kedosha Janina, New York.

11. UK Data Archive, SN 4732, European Social Survey, 2002–03; "On European Attitudes toward Israel, Iran, and North Korea," *ParaPundit*, November 2003.

12. Yuval Yoaz, "State Refuses to Register 'Israeli' Nationality," *Ha-Aretz* (English ed.), May 19, 2004.

13. In summer 2001, the Greek and foreign press were full of the story. For a summary, see Takis Michas, "A Greek Tempest in a Teapot Boils Over," Dow Jones News Service, August 28, 2001.

14. Cited in ibid.

15. "The National Commission for Human Rights met on 13 July 2000 and discussed the issue of the recording of the religion of the holder on police identity cards, which it regarded as falling within its competence in accordance with Law 2667/1998. . . . By way of conclusion, both the compulsory and optional re-

cording of the holder's religion on identity cards is unconstitutional (Article 5, paras. 1 and 2, Article 13 of the Constitution) and contrary to Greek legislation (Convention of the Council of Europe) on the protection of the individual from the automated processing of information of a personal character—ratified by Law 2068/1992, Directive 95/46 of the European Parliament and Council on the protection of natural persons from the processing of data of a personal character, Articles 18, 26 and 27 of the International Covenant on Civil and Political Rights (ratified by Law 2462/1997), Articles 9 and 14 of the European Convention on Human Rights (ratified by Presidential Decree 53/1974), Article 18 of the Universal Declaration of Human Rights (1948), and the Declaration of the United Nations on the elimination of all forms of intolerance and discrimination based on religion or belief (1981)." Press Release, Hellenic Republic, National Commission for Human Rights, July 13, 2000, Athens.

INDEX